# FOUNDATIONS OF STREAM AND RIVER ECOLOGY

# FOUNDATIONS OF STREAM AND RIVER ECOLOGY

## A GUIDE TO THE CLASSIC LITERATURE

Edited by Wyatt F. Cross,
Jonathan P. Benstead,
Amy M. Marcarelli, and
Ryan A. Sponseller

THE UNIVERSITY OF CHICAGO PRESS
*Chicago and London*

The University of Chicago Press, Chicago 60637
The University of Chicago Press, Ltd., London

Published 2024
Printed in the United States of America

33 32 31 30 29 28 27 26 25 24    1 2 3 4 5

ISBN-13: 978-0-226-83711-6 (cloth)
ISBN-13: 978-0-226-83713-0 (paper)
ISBN-13: 978-0-226-83712-3 (e-book)
DOI: https://doi.org/10.7208/chicago/9780226837123.001.0001

Library of Congress Cataloging-in-Publication Data

Names: Cross, Wyatt F., editor. | Benstead, Jonathan P., editor. | Marcarelli, Amy M., editor. | Sponseller, Ryan A., editor.
Title: Foundations of stream and river ecology : a guide to the classic literature / edited by Wyatt F. Cross, Jonathan P. Benstead, Amy M. Marcarelli, and Ryan A. Sponseller.
Description: Chicago ; London : The University of Chicago Press, 2024. | Includes bibliographical references and index.
Identifiers: LCCN 2024016397 | ISBN 9780226837116 (cloth) | ISBN 9780226837130 (paperback) | ISBN 9780226837123 (ebook)
Subjects: LCSH: Stream ecology. | Environmental literature.
Classification: LCC QH541.5.S7 F68 2024 | DDC 577.6/4—dc23/eng/20240508
LC record available at https://lccn.loc.gov/2024016397

♾ This paper meets the requirements of ANSI/NISO Z39.48-1992 (Permanence of Paper).

DEDICATED TO OUR MENTORS AND STUDENTS,
PAST, PRESENT, AND FUTURE.

# CONTENTS

## 4. SPECIES INTERACTIONS AND COMMUNITY STRUCTURE 45

*Angus R. McIntosh and Jonathan D. Tonkin*



## 5. FOOD WEBS AND TROPHIC INTERACTIONS 61

*Alexander D. Huryn and Alan G. Hildrew*

## 8. NUTRIENT DYNAMICS 107

*Daniel von Schiller, Eugènia Martí, Marc Peipoch, and H. Maurice Valett*

Meyer, J. L., and G. E. Likens. 1979. "Transport and Transformation of Phosphorus in a Forest Stream Ecosystem." *Ecology* 60:1255–69. 109

Webster, J. R., and B. C. Patten. 1979. "Effects of Watershed Perturbation on Stream Potassium and Calcium Dynamics." *Ecological Monographs* 49:51–72. 110

Newbold, J. D., J. W. Elwood, R. V. O'Neill, and W. Van Winkle. 1981. "Measuring Nutrient Spiralling in Streams." *Canadian Journal of Fisheries and Aquatic Sciences* 38:860–63. 112

Newbold, J. D., J. W. Elwood, R. V. O'Neill, and A. L. Sheldon. 1983. "Phosphorus Dynamics in a Woodland Stream Ecosystem: A Study of Nutrient Spiralling." *Ecology* 64:1249–65. 112

Grimm, N. B. 1987. "Nitrogen Dynamics during Succession in a Desert Stream." *Ecology* 68:1157–70. 114

Triska, F. J., V. C. Kennedy, R. J. Avanzino, G. W. Zellweger, and K. E. Bencala. 1989a. "Retention and Transport of Nutrients in a Third-Order Stream: Channel Processes." *Ecology* 70:1877–92. 115

Triska, F. J., V. C. Kennedy, R. J. Avanzino, G. W. Zellweger, and K. E. Bencala. 1989b. "Retention and Transport of Nutrients in a Third-Order Stream in Northwestern California: Hyporheic Processes." *Ecology* 70:1893–905. 115

Peterson, B. J., W. M. Wollheim, P. J. Mulholland, J. R. Webster, J. L. Meyer, J. L. Tank, E. Martí, W. B. Bowden, H. M. Valett, A. E. Hershey, W. H. McDowell, W. K. Dodds, S. K. Hamilton, S. Gregory, and D. D. Morrall. 2001. "Control of Nitrogen Export from Watersheds by Headwater Streams." *Science* 292:86–90. 118

## 9. RECIPROCAL STREAM-LAND INTERACTIONS 123

*Colden V. Baxter, Amanda Subalusky, and Hiromi Uno*

Sedell, J. R., and J. L. Froggatt. 1984. "Importance of Streamside Forests to Large Rivers: The Isolation of the Willamette River, Oregon, USA, from Its Floodplain by Snagging and Streamside Forest Removal." *Internationale Vereinigung für theoretische und angewandte Limnologie: Verhandlungen* 22:1828–34. 125

Naiman, R. J., J. M. Melillo, and J. E. Hobbie. 1986. "Ecosystem Alteration of Boreal Forest Streams by Beaver (*Castor canadensis*)." *Ecology* 67:1254–69. 127

Junk, W. J., P. B. Bayley, and R. E. Sparks. 1989. "The Flood Pulse Concept in River-Floodplain Systems." *Canadian Journal of Fisheries and Aquatic Sciences* Special Publication 106:110–27. 128

# FOREWORD

When I took my comprehensive exam as a doctoral student in 1981, I figured I had found and read all the relevant literature in my field of stream ecology. I was likely wrong, but such a feat was possible in those days: there were "key" journals and soon-to-be-classic papers that cited other classic papers, and one day a week in the library's "current periodicals" reading room was all you needed. Today, most searching is done online and spans a bewildering number of journals. It is tempting to set the parameters of your search to pick up only the last few years (and many students do), meaning that our classic, foundational papers of stream ecology are missed. This book is a gift to all of us who want our students (and ourselves) to acknowledge and continue to learn from those building blocks.

In our dinner and discussion group at ASU, the idea of assembling a compendium of classics in stream ecology has come up several times (with several lists buried in paper files), but such a collection never materialized. It was, and is, an important contribution to make, and I want to thank the editors for doing so. Of course, my group's idea of what constitutes a classic paper is likely different from your idea or the ideas of this book's editors and contributing authors. Slobodkin (1992) went so far as to say, "Each generation sees as important the things that they imagine have led to themselves." But there's little of that self-promotion evident in the six or seven selected articles in each section of this book, to the credit of the editors and the outstanding teams of contributors they recruited to the project.

Cross, Benstead, Marcarelli, and Sponseller have succeeded in putting together a book that will help current and future generations of stream ecologists understand the roots of the evolving field of stream ecology. The introduction, article commentaries, and prospects for future research presented in each of the ten book chapters are enjoyable to read, and they capture the essence of papers that span the decades from the late 1960s to the early twenty-first century. Anyone who reads through this book will come away with a good overview of the most influential literature in the field, but perhaps even more importantly, it will spark motivation to answer as-yet-unresearched questions starting from a solid footing.

For me, reading *Foundations of Stream and River Ecology* was comforting, nostalgic, and confirmatory (maybe I had read all the important literature!). It also reminded me of a defining feature of this field of stream ecology—not so much in terms of the science but in terms of the community of scientists. We aren't siloed. We share a fascination for these ecosystems and curiosity that extends beyond our own specific areas of expertise. For those who are starting out in stream ecology—in any area covered by this collection or even in areas not yet envisioned—this book is an excellent first step into the stream world. And all of us will benefit from discovering, or rediscovering, the foundations that support the expansion of our collective knowledge in stream and river ecology.

**NANCY B. GRIMM**
Arizona State University

# PREFACE

It is exciting and uniquely challenging to build a book about the foundations of one's field. This book represents years of deep thinking about stream ecology, countless classroom sessions and reading groups, and many discussions about literature with students, friends, and colleagues, all motivated by a shared enthusiasm about streams and a desire to better understand and protect them. Pulling together a list of classics to include in this book was an opportunity to revisit the papers that inspired us throughout our careers, and then to reread them, compare them, and debate their influence on the development and future of our field. It also allowed us to reflect on the value of historical context for students just getting started on their professional journeys.

The book is divided into ten topical chapters that together provide a broad overview of the foundations of stream and river ecology, covering aspects of the physical environment, biological communities and interactions, ecosystem processes, stream-land connections, and the importance of place-based research. Although we have tried our best to highlight the most important contributions, there are far too many to discuss in a book of this length, and many important papers have necessarily been omitted. You may be familiar with other *Foundations* books, such as the 1991 *Foundations of Ecology: Classic Papers with Commentaries*, edited by Leslie A. Real and James H. Brown. If so, you'll notice some differences in our volume. The highlighted papers are not reprinted in full. In lieu of reprints, each chapter contains an introductory section, brief summaries and contextualization of the highlighted papers, and a forward-looking conclusion. Where appropriate, we cross-reference papers from other chapters, and we boldface citations of highlighted papers throughout the entire volume. Although we regret that the full text of some highlighted papers may be difficult to access, we think readers will appreciate this alternative approach, as it affords a greater level of context and insight about each paper, as well as additional citations and related content that provide access points for those interested in a deeper dive into each topic.

Stream ecology as a field continues to grow and become ever more collaborative and interdisciplinary. In the past few decades, many of the core ideas highlighted in this volume have been expanded on, revised, and pushed in new, and perhaps unforeseen, directions that provide a better understanding of how streams function. We have also witnessed a substantial increase in the number and pace of published studies about streams, and although we have new ways of searching and synthesizing the literature, it is still easy to feel overwhelmed or to lose sight of where we have been and where we are headed. Indeed, we produced this volume, in large part, to provide an entry point to the literature that is anchored by its most foundational contributions, but also to encourage a deep reading (or rereading) of these classic papers, which still have much to teach us and new ideas to inspire. In our view, identifying the critical problems and knowledge gaps of tomorrow requires that we remember how core ideas in stream ecology originated and how they have changed and matured over time. We hope this book will be an inspiration for current and future students, introduce them to groundbreaking work of the past, and catalyze new and exciting research in streams and rivers that will continue to expand the boundaries of our field.

Finally, in planning this volume, we recognized that identifying foundational papers is a human process, influenced by the identities of the editors and chapter authors. Our respective backgrounds and perspectives influence the papers to which we were introduced and now value as foundational contributions to the field. Our identities also dictate the ways that we have experienced our field, first as students and now as scientists and professors. Thus, building this collection required us to reconcile the task of selecting truly foundational papers with the social context in which this work was produced, including the undoubted advantages enjoyed by some investigators. The simple fact is that our field was historically dominated by white, male scientists of European descent. We (the editors) acknowledge that we are included in these groups, and that we were trained in the United States academic system. We therefore worked to be as inclusive as possible in our selection of papers and chapter authors, but there is clearly much work to be done with respect to increasing diversity, equity, and inclusion in the field of stream ecology. There is similarly a rich history of literature published in French, German, and Russian, which is generally not featured here and was last synthesized in Noel Hynes's foundational 1970 textbook *The Ecology of Running Waters*. We celebrate ongoing efforts to diversify the field by supporting and building international connectivity, especially with scientists from Africa, Latin America, and other historically underrepresented regions, as well as continued efforts to achieve equity by supporting women, BIPOC, and LGBTQIA+ scientists across all career stages in our labs, field sites, and societies. We are hopeful that the future of stream ecology will not look like its past.

**WYATT F. CROSS, JONATHAN P. BENSTEAD,**
**AMY M. MARCARELLI, AND RYAN A. SPONSELLER**
November 30, 2023

# 1

# ORIGINS AND ANTECEDENTS

## THE HEADWATERS OF STREAM ECOLOGY

Jonathan P. Benstead, Amy M. Marcarelli,
Ryan A. Sponseller, and Wyatt F. Cross

Stream ecology is a young science, having only emerged as a distinct discipline in the mid-twentieth century as approaches rooted in aquatic entomology, fisheries biology, water pollution management, and traditional limnology coalesced with the maturing ecosystem concept to form a new perspective on streams as unique ecosystems. This chapter provides a brief history of the early development of stream ecology and highlights seven foundational papers that appeared during this period when stream ecology was emerging as a nascent field and before the first textbook devoted to the ecology of running waters (Hynes 1970) had been published. Our chosen papers provide a window into an exciting era of early discovery and foundation building that would develop into the rich and diverse field of modern stream ecology. Although inevitably subjective to some extent, our choices were also guided by an explicit focus on pre-1970 publications and care not to preempt the selection of foundational papers by the authors of the thematic chapters to follow.

An underappreciated fact today is that much of contemporary stream ecology has its origin in applied questions motivated by the management of recreational riverine fisheries and water quality. Stephen Forbes's (1903) early research at the Illinois Natural History Survey on the food of fish and Kolkwitz and Marsson's (1902) invertebrate-based stream bioassessment scheme remind us of this fact. Many early studies on the taxonomy and distributions of stream insects were conducted in the context of their roles as food for game fish (e.g., Dodds and Hisaw 1925;

Carpenter 1927; Whitehead 1935; Needham 1938; Ide 1942). This focus on invertebrates as prey for fish explains the otherwise peculiar emphasis of aquatic entomology on immature life-history stages, in contrast to terrestrial entomology. As knowledge about stream invertebrates grew, stream communities were increasingly studied for their own sake (Davidson and Wilding 1943; Sprules 1947; Berg 1948; Badcock 1949). At the same time, studies began to move beyond description of basic patterns to focus increasingly on the ecological relationships among invertebrates, fish, and their physical environment (Linduska 1942; Pennak and Van Gerpen 1947; Jones 1949). The recognition that different taxa had different tolerances to sewage and industrial pollution was particularly important, especially for understanding the relationship between fisheries management and stream condition (Richardson 1921; Ellis 1937; **Patrick 1949**).

The 1950s saw the advent of increasingly sophisticated approaches, not just in describing patterns but also in analyzing processes in streams. Some studies of this era have become classics (Berner 1951; Jones 1951; Scott 1958; Armitage 1958). Although many studies continued to be motivated by fisheries management (e.g., Ellis and Gowing 1957; Huet 1959), these efforts had become more ambitious, as exemplified by K. R. Allen's (1951) monumental monograph on the ecology of brown trout in the Horokiwi Stream in New Zealand (see chapter 5). This decade also saw the first ecosystem-level studies in streams as pioneering ecologists became interested in the flow of energy through aquatic ecosystems (e.g., **Odum 1956** [chapter 7], 1957; Teal 1957). Finally, this period saw active and continuing methods development, as sampling techniques were adapted from other fields and novel solutions were developed in response to the unique challenges of collecting meaningful data in flowing water, especially in view of the high spatial and temporal variability that characterizes streams, including variation in the distribution of benthic invertebrates (Needham and Usinger 1956; Macan 1958).

The 1960s witnessed an explosion of research on stream communities and ecosystems as both methodologies and taxonomic knowledge improved. This was the golden age of pioneering, detailed community-level investigations, including those by **Hynes (1961)** and **Minckley (1963)**. Studies also began to tackle increasingly complex topics, including the physicochemical drivers of community structure (Edington

1968; Harrel and Dorris 1968; Cummins and Lauff 1969) and the longitudinal zonation of rivers in relation to patterns of community structure (Kuehne 1962; Illies and Botosaneanu 1963; Sheldon 1968). This period also saw increased attention on energy flow (Warren et al. 1964; McIntire and Phinney 1965; Duffer and Dorris 1966; Wright and Mills 1967; Tilly 1968) and the first comprehensive study of nutrient cycling in a stream, in what must be the first appearance of the word "ecosystem" in the title of a stream ecology publication (**Ball and Hooper 1963**).

Particularly notable during the 1960s was an increased emphasis on the trophic structure of stream communities. Representative studies range from detailed, taxon-based descriptions of gut contents (Brown 1961; Mecom and Cummins 1964; Mann 1965) to analyses of whole-stream food webs (Chapman and Demory 1963; Chapman 1966; **Cummins et al. 1966** [chapter 5]). These approaches informed a new appreciation of the energetic importance of allochthonous detritus, paving the way for what would eventually become a large body of literature on the dynamics of terrestrial organic matter imported to stream channels (**Nelson and Scott 1962**; Hynes 1963; Egglishaw 1964; **Minshall 1967** [chapter 5]; Kaushik and Hynes 1968; Mathews and Kowalczewski 1969).

Also significant during this period was the growth of studies devoted to invertebrate drift (Waters 1961, 1965; Elliot 1967; Anderson and Lehmkuhl 1968; see chapter 4). These studies can be seen as both a throwback to the roots of stream ecology in fisheries management and a reflection of the field's increased emphasis on understanding the population ecology and life history of stream invertebrates (Waters 1966; Bailey 1966; Dimond 1967; Holt and Waters 1967). The decade preceding the publication of *The Ecology of Running Waters* was also marked by renewed interest in the effects of pollution in streams, synergized by Hynes's (1960) volume on the topic, including the consequences of pesticide use to control nuisance aquatic species (Hynes and Williams 1962; Brinkhurst 1965; King and Ball 1967). Finally, the 1960s saw the appearance of the first review papers in stream ecology as authors attempted to synthesize the burgeoning literature on the topic and to catalyze new ways of conceptualizing stream ecosystems (**Margalef 1960**; Macan 1962; Cummins 1966). Thus, by the end of the 1960s, stream ecology had emerged as a distinct field of study.

## HIGHLIGHTED PAPERS

Patrick, R. 1949. "A Proposed Biological Measure of Stream Conditions, Based on a Survey of the Conestoga Basin, Lancaster County, Pennsylvania." *Proceedings of the Academy of Natural Sciences of Philadelphia* 101:277–341.

Although Patrick's 1949 paper was preceded by early European literature (Kolkwitz and Marsson 1902), the maturation of comprehensive stream bioassessment is clearly evident in this paper, which was motivated by rapid changes in land use and waterways that were threatening the quality of freshwater in the United States. This paper is one of Ruth Patrick's earliest foundational contributions, which later earned her the National Medal of Science and induction into the National Women's Hall of Fame. The objective of this study was to determine whether changes in diversity within different taxonomic groups could be used as a measure of biotic condition in streams, accounting for the variability in their responses to, and tolerance of, physical and chemical factors. She tested these ideas in the Conestoga Basin, in Pennsylvania, USA, where water quality was degraded to different degrees by industrial waste and sewage. Her study also built on previous ideas in the field, including the notion of assimilative capacity, referring to the fact that ecosystems show resilience to pollution below some critical threshold, and the use of organisms as indicators based on their tolerance of environmental conditions (Kolkwitz and Marsson 1902). The novel contribution of Patrick and her team was to use the entire "biodynamic system"—from bacteria to fish—to detect responses to water-quality degradation within a catchment.

The survey was ambitious in both its sampling and spatial scope. It encompassed fifty-nine sites across seven streams in a 1,230 $km^2$ catchment, and involved the characterization of water chemistry, isolation of water and sediment bacteria, and the cataloging of algae, flowering plants, bryozoans, zooplankton, benthic invertebrates, and fish at each site. These data were combined to classify sites as "healthy," "semi-healthy," "polluted," or "very polluted." The general effect of pollution was a reduction in species numbers and increased dominance of tolerant species in impacted sites, even when chemical and physical factors did not clearly indicate degraded conditions. This finding led Patrick to conclude that assessing biological community composition was far superior to monitoring chemical and physical variables because concentrations of solutes vary widely over time, and because field measurements cannot discern interactions among chemical and physical factors that have detrimental effects on organisms.

The key contribution of Patrick's paper was to establish biological assessment as a fundamental way to evaluate surface-water pollution (Karr 1991), which was later dubbed "the Patrick principle." Biotic integrity was thereafter encoded as an essential characteristic of water bodies, equal in importance to chemical and physical conditions, in Section 101 of the 1972 US Clean Water Act, which Patrick helped to author (Lowe 2015). The challenge of classifying biotic integrity using integrated observations and modeling of species distributions still underpins techniques used to monitor and assess water quality in streams, rivers, and lakes globally (Karr 1991; Norris and Hawkins 2000).

Margalef, R. 1960. "Ideas for a Synthetic Approach to the Ecology of Running Waters." *Internationale Revue der gesamten Hydrobiologie und Hydrographie* 45:133–53.

Throughout much of its history, stream ecology has focused on understanding patterns and processes in running waters, rather than using these ecosystems to advance ecological theory (Fisher 1997). As Spain's first professor of ecology, Ramón Margalef stood out as an early exception to this rule in that he was keenly interested in developing unifying principles that applied across ecological systems (e.g., Margalef 1963). In his thought-provoking essay published in 1960, he presented a wide range of hypotheses to explain ecological patterns in river ecosystems, developed in the context of prominent and general ecological themes, such as ecosystem structure, productivity, and succession.

Margalef's essay is organized into three sections, which address the structure and dynamics of populations, the application of successional concepts to running waters, and a classification scheme for describing longitudinal patterns of riverine communities. Following an introduction to the basic structural elements of rivers, he presents a conceptual framework for understanding the distribution of benthic and pelagic species along watercourses that is anchored by the fundamental relationships among biomass, productivity, and downstream movement. His formulation emphasizes the varied importance of productivity in maintaining populations given downstream changes in hydrology, as well as the differences among species in their susceptibility to losses through transport and/or sinking. Emerging from this view are more synthetic ideas related to the longitudinal organization of riverine communities, including the overwhelming importance of downstream transport of living and nonliving materials. Margalef also draws parallels between community change along rivers and ecological succession, arguing that longitudinal patterns in the collective properties of communities can be viewed as a spatial manifestation of ecosystem development or maturation. Using these ideas as a guide, he concludes by presenting a classification system for stream algal communities, identifying sets of taxa that are likely to prosper along gradients of ecosystem maturation.

Margalef's essay is replete with innovative ideas, most ahead of their time, several that helped develop concepts that would fully emerge in the 1970s and 1980s, and others that would be challenged. His general ideas on ecological succession, for example, would be featured in Odum's (1969) classic treatise on this topic. His more specific predictions related to ecosystem maturity were fodder for discussion in several landmark studies in streams (e.g., **Minckley 1963**), including those that advanced the ecosystem concept (e.g., **Fisher and Likens 1973** [chapter 7]) and the critical role of physical disturbance and recovery (Fisher et al. 1982). Although his arguments for longitudinal succession were dismissed in the river continuum concept (**Vannote et al. 1980** [chapter 3]), his ideas related to upstream-downstream connectivity nonetheless lie at the core of that concept. Finally, Margalef's view of running waters was, at the time, notably holistic, in that understanding streams as ecosystems requires explicit consideration of the dynamics shaping communities, pools and fluxes of organic matter, upstream-downstream connections, and even lateral connections to catchment soils. That this perspective now seems so obvious to stream ecologists owes much to Margalef's vision.

Hynes, H. B. N. 1961. "The Invertebrate Fauna of a Welsh Mountain Stream." *Archiv für Hydrobiologie* 57:344–88.

Noel Hynes's 1961 publication, and much of his earlier work, epitomizes the convergence of multiple disciplinary "tributaries"—invertebrate zoology and entomology, taxonomy, and natural history—that contributed to defining stream ecology as a discipline in its own right. In the mid-1950s, Hynes became increasingly aware that autecological approaches would be insufficient for understanding the effects of pollution on streams (Hynes 1960), or for revealing how and why stream communities change over time (Hynes 1981). He thus began one of the earliest comprehensive studies of a stream invertebrate community, in the Afon Hirnant, a relatively undisturbed Welsh mountain stream. His study involved monthly quantitative sampling of the benthos over an annual cycle (1955–1956) and long hours in the laboratory separating, counting, and identifying individuals. By paying close attention to the full range of invertebrate taxa, body sizes, and life stages, Hynes provided one of the first annualized characterizations of life cycles and trophic structure for an entire stream community.

Hynes's 1961 paper includes a clever visualization of the timing of growth and interrelationships among the many coexisting invertebrate taxa (figure 1.1). Hynes used this information to categorize different life histories (i.e., seasonal, slow seasonal, fast seasonal), an approach that inspired many subsequent studies on the life cycles and production dynamics of stream invertebrates (e.g., Huryn and Wallace 2019). He also leveraged the limited knowledge of invertebrate feeding relationships available at the time to speculate about the coincidence of predator life histories with those of their prey. Finally, Hynes took the bold, and similarly speculative, step of estimating community secondary production by summarizing the death rates of a community-averaged "cohort" of individuals. This step marked the beginnings of a new quantitative method for estimating energy flux through stream communities, refining the cruder approaches of earlier studies (Odum 1957; Teal 1957) and setting the stage for detailed investigations that would place streams at the forefront of efforts to determine animal secondary production in ecosystems (Benke 2010; Benke and Huryn 2017). If we consider this paper alongside his many other contributions to the field, it is safe to say

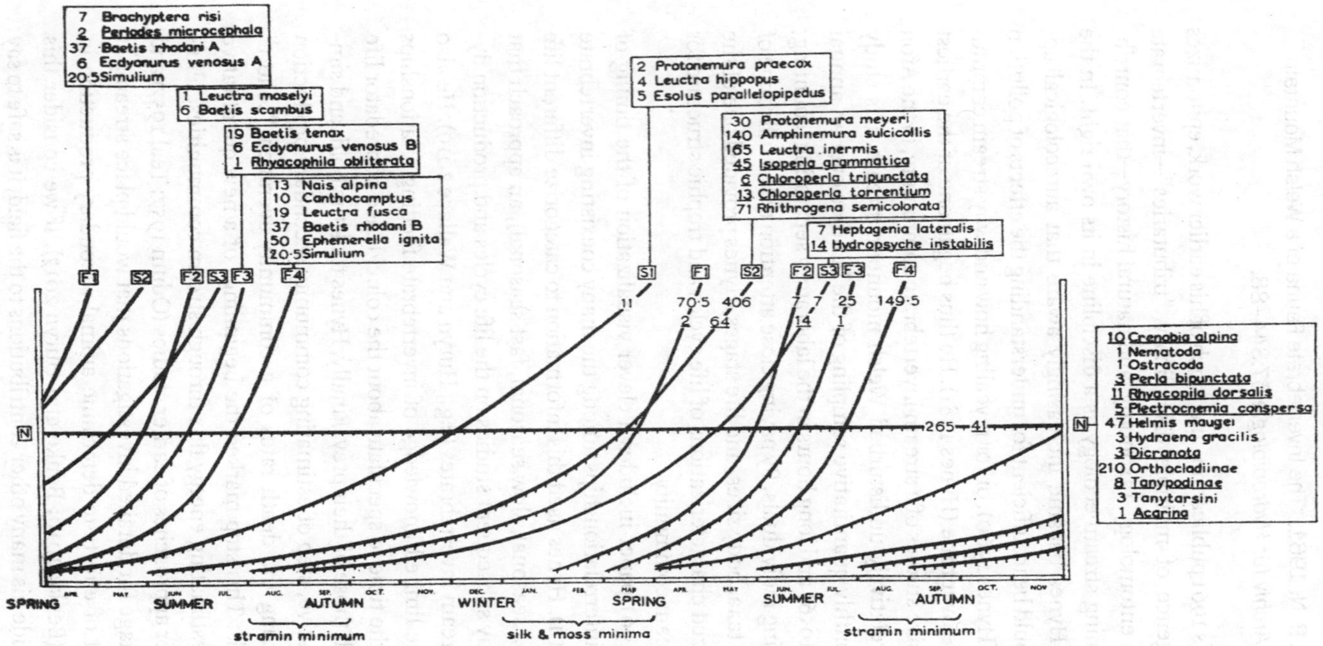


Fig. 1.1 Diagram illustrating the types of life cycle, and their interrelationships, in the invertebrate community of stations 1–4 in the Afon Hirnant. The vertical axis indicates increase in size. The figures are the "index" numbers of the species and of the various types of life cycle. The names of the carnivores and their index numbers are underlined. Periods during which hatching eggs are causing continuous recruitment to the various cycles are shown by short vertical hatching. Hynes, H. B. N. 1961. "The Invertebrate Fauna of a Welsh Mountain Stream." *Archiv für Hydrobiologie* 57:344–88.

that Hynes may be the grandfather of what he would have called "lotic limnology" (Hynes 2001).

Nelson, D. J., and D. C. Scott. 1962. "Role of Detritus in the Productivity of a Rock Outcrop Community of a Piedmont Stream." *Limnology and Oceanography* 7:396–413.

Nelson and Scott determined the trophic structure and productivity of a river invertebrate community, providing, along with **Hynes (1961)**, one of the earliest estimates of secondary production at the whole-community level. Their focus was on energy sources and the productivity of invertebrates inhabiting a distinctive stream habitat—bedrock outcrops, or "shoals"—that is common in the geologically complex Piedmont plateau of the eastern United States. Their fieldwork took place between 1954 and 1957 in a single area of shoal habitat in the Middle Oconee River in Georgia, USA, where shallow bedrock was colonized by a firmly attached aquatic plant, *Podostemum ceratophyllum*.

Nelson and Scott quantified the production of biotic stream compartments by tracking net changes in biomass over successive sampling periods. The resulting estimate, dubbed "net apparent production," is a crude measure of biomass production, including that of invertebrates, because it ignores losses due to drift, emergence, or consumption. Nevertheless, the study was one of the first in streams to use production estimates to calculate ecosystem-level metrics such as biomass turnover and production efficiencies. Another unusual feature of the study was the size of the stream, which was sixth-order and had a mean discharge of 8 $m^3/s$. Even today, little is known about the energetics (and other aspects) of such mid-order streams compared with headwaters, underlining the pioneering character of the study. Grubaugh and Wallace (1995) resampled the site after thirty-four years of land-use change (from mostly row-crop agriculture to forest) and associated increases in water quality, which also makes Nelson and Scott's study a rare benchmark for characterizing long-term changes in stream ecosystems.

Nelson and Scott's data showed little evidence for seasonality. Instead, biomasses of most taxa were greatly affected by discharge variation. Both scouring during storm flows and desiccation during periods of low flow proved important. Thus, the study provided early evidence for the overwhelming importance of flow disturbance in structuring stream and river communities, many years before disturbance became a mainstream interest in stream ecology. In addition, the results pointed to detritus as the primary energy source for benthic invertebrates and the driver of the heterotrophic state of the stream, a conclusion derived from the comprehensive analysis of potential organic matter sources, which introduced a perspective that was new at the time. Finally, the paper attempted to assign the collected invertebrates to feeding groups, an early effort that eventually led to the development of the different and widely adopted functional feeding group concept (Cummins 1973). However, Nelson and Scott struggled with the implications of omnivory for energy flow, especially with regard to the importance of microbes as energetic links between detritus and detritivores. All these topics developed into major themes of inquiry in stream ecology in the ensuing decades.

Ball, R. C., and F. F. Hooper. 1963. "Translocation of Phosphorus in a Trout Stream Ecosystem." In *Radioecology*, edited by V. Schultz and A. W. J. Klement, 217–28. Washington, DC: Reinhold Publishing Corporation and the American Institute of Biological Sciences.

Ball and Hooper took a groundbreaking approach to studying nutrient cycling and trophic structure in stream ecosystems. A recommended precursor to their 1963 paper is Aldo Leopold's essay "Odyssey of Atom X," first published in 1942 by *Audubon* magazine and reprinted in *A Sand County Almanac* (Leopold 1949). "Odyssey," which describes the centuries-long journey of a single atom through biotic and abiotic pools from the prairie uplands of the United States to the ocean, is a portrayal of nutrient cycling written at a time when the contributing processes could not be studied with any precision in natural ecosystems. It was not until radioisotopes made highly sensitive tracer approaches possible that the movement of elements through ecosystems could be tracked in a quantitative fashion. Although radioisotopes had been adopted earlier for studying elemental cycling in lakes (Hutchinson and Bowen 1947), Ball and Hooper were the first to track the experimental addition of a radioisotope within a stream ecosystem, where the fates of elements differ fundamentally from those in lakes because elemental cycling is superimposed onto directional downstream transport.

Ball and Hooper carried out their study in the West Branch of the Sturgeon River in Michigan, USA, a cold-water stream characterized by low ambient phosphorus (P) concentrations (~1 μg/L) and low productivity. After four summers (1954–1957) of recording baseline data on P distribution in major stream compartments within a 4 km reach, they added $^{32}$P-labeled phosphoric acid for 33 min in each of the following three summers (1958–1960) to track the radiolabel in water, periphyton, aquatic macrophytes, macroinvertebrates, and fish over several weeks. The calculated P uptake rate differed greatly among compartments, sub-reaches, and years. Reflecting Ball and Hooper's ecosystem perspective, their paper formalized the "average distance" traveled by a nutrient atom before being taken up, which we now call uptake length, and documented the importance of transient storage and dispersion during advective downstream transport. Furthermore, the pattern of $^{32}$P uptake varied notably among animal consumers, which could be attributed to differences in feeding mode, body size, and metabolic turnover rate. Thus, this paper was the first to link nutrient cycling explicitly to the trophic structure of communities in stream ecosystems.

Given that the analysis of isotope data was still in its infancy in the 1960s, it is not surprising that the interpretation of the paper's results partially relies on conjecture. Nevertheless, many of the topics it raised would later become major research foci in stream ecology. In particular, Ball and Hooper preceded the conceptualization of nutrient spiraling by at least a decade. Furthermore, their study inspired the further use of radioactive P isotopes, notably in a series of other classic studies carried out in Walker Branch, Tennessee, USA (e.g., Elwood and Nelson 1972; **Newbold, Elwood, O'Neill et al. 1983** [chapter 8]; Mulholland et al. 1985), and informed numerous other contributions such as **Webster and Patten (1979** [chapter 8]), **Newbold et al. (1981** [chapter 8]), and Elwood et al. (1983), culminating in the Stream Solute Workshop (1990), which formalized a suite of metrics that are now routinely used to describe the "spiraling" of elements in stream ecosystems. Finally, Ball and Hooper laid important groundwork for later constructions of stream phosphorus budgets, such as **Meyer and Likens (1979** [chapter 8]), as well as for the development of stable-isotope techniques to trace elements in ecosystems without the environmental and health risks associated with the release of radioisotopes (e.g., **Peterson et al. 2001** [chapter 8]).

Minckley, W. L. 1963. "The Ecology of a Spring Stream, Doe Run, Meade County, Kentucky." *Wildlife Monographs* 11:3–124.

Minckley's 1963 paper is an encyclopedic tour de force, cataloging the physical, chemical, and biological characteristics, as well as their interactions, in a single stream, Doe Run, in northern Kentucky, USA. Minckley's goals were (1) to characterize in exhaustive detail an unpolluted spring-fed stream before the downstream half was impounded, and (2) to provide data to advance understanding of ecological structures and processes. Amazingly, this extensive monograph was only part of Minckley's PhD dissertation in biology and geology, the influence of which is evident in the detailed and integrated analysis of this spring-fed stream.

The paper starts with documentation of the physical setting of the stream and springs in a region of extensive karst topography, based on five main sampling stations arrayed between the stream's spring origin and its confluence with the Ohio River. Minckley then presents qualitative and, where possible, quantitative surveys and species inventories of algae, plants, invertebrates, and fish at each site, including descriptions of interactions among species and between the species and the physical environment. Next, he describes diets of fishes and invertebrates and uses those to build a food-web map showing the trophic structure of primary and secondary consumers, highlighting the role of omnivory and the contributions of detritus to the diets of both invertebrates and fishes. Minckley then integrates his observations into a discussion on the roles of physical disturbance versus biological succession in structuring stream communities, and proposes that streams are "biological and chemical individuals" that experience unique climate and weather. He sets forth a framework for characterizing invertebrate communities on the basis of physical habitat conditions and species associations, anticipating the later foundational contribution on species traits in rivers by **Townsend and Hildrew (1994** [chapter 3]). The paper's conclusion lays out a sophisticated understanding of the ways that stream communities are structured by multiple factors interacting across spatial scales, as has now become common knowledge in stream ecology.

Minckley's monograph is rich in data, novel observations, and synthetic ideas that are antecedent to many different lines of research in the pantheon of stream ecology. This is evident in the diversity of ways that the paper has been cited in the literature. For example, it is commonly recognized as one of the earliest demonstrations of the importance of allochthonous organic matter for stream consumers and ecosystems (e.g., **Minshall 1967** [chapter 5]; Triska et al. 1984), which is notable because Minckley emphasizes the importance of detritus from both autochthonous and allochthonous sources, as was later pointed out by **Minshall (1978** [chapter 7]). Similarly, Minckley's comprehensive account of the population dynamics and feeding ecology of benthic stream invertebrates filled knowledge gaps in the ecology of these taxa in North America (Rosemary J. Mackay, personal communication), as reflected by citations referring to information on invertebrate natural history (e.g., Cummins 1973; **Webster and Patten 1979** [chapter 8]). However, perhaps its most important contribution was a conceptual framework for the importance of physical versus biological drivers in structuring stream communities, thus directly challenging the model of stream succession proposed by **Margalef (1960)**. The idea that physical setting and disturbance could control or reset biological trajectories influenced the river continuum concept (**Vannote et al. 1980** [chapter 3]; Minshall et al. 1983), as well as complementary models describing the dynamics of streams where climatic conditions lead to highly unpredictable disturbances (e.g., Winterbourn et al. 1981; Fisher et al. 1982).

Ross, H. H. 1963. "Stream Communities and Terrestrial Biomes." *Archiv für Hydrobiologie* 59:235–42.

Ross was among the first to suggest that terrestrial riparian and stream biota shared an evolutionary history, hypothesizing that riparian vegetation and climate both played prominent roles in structuring stream communities. His 1963 paper was an early attempt to go beyond local descriptions of communities and to consider both evolutionary and ecological factors as drivers of biogeographic patterns. Leveraging his vast knowledge of aquatic entomology and evolution, Ross hypothesized a relationship between stream insect communities and terrestrial biomes. Specifically, he observed distinct geographic differences in caddisfly communities that appeared to follow the boundaries of major biomes in North America, including eastern deciduous forest, western montane forest, and semidesert scrub. In addition, he found that these differences were most pronounced among small streams, where the direct connection between terrestrial and aquatic ecosystems is greatest. For the eastern deciduous forest biome, for example, Ross suggested that both ecological and climatic characteristics, including dense shade for much of the year, reduced shade during the cold season, large pulses of leaf-litter inputs to stream channels, and a characteristic precipitation and thermal regime, combine to determine survival, composition, and local evolution and diversification of Trichoptera.

Although Ross's paper was short and in parts speculative, it had a large influence on the development of stream ecology. Its ideas were taken up both in the river continuum concept (**Vannote et al. 1980** [chapter 3]) and in **Hynes's (1975** [chapter 2]) visionary perspective on "the stream and its valley" in making connections between river insect communities and riparian vegetation. Furthermore, Ross was acknowledged in a series of other influential publications (e.g., Cummins 1973, 1974; Minshall et al. 1983; Cummins et al. 1989) that began to build a comprehensive conceptual basis for stream trophic and ecosystem ecology. Remarkably, at the same time, his paper highlighted the value of taxonomy and phylogenetics in contributing to our understanding of community assembly and ecological interactions in streams (Wiggins and Mackay 1978).

## LOOKING FORWARD

Only a few aquatic scientists would have self-identified as stream ecologists prior to 1970. The decades since have seen tremendous growth. Pioneering studies like the seven papers we have selected here were instrumental in defining stream ecology as a field and thus setting the stage for a new and vibrant discipline. Importantly, we suggest not viewing these papers, or those presented in subsequent chapters, as mere historical footnotes that have been superseded by more recent work. Instead, these studies have all stood the test of time and continue being worthy of fresh consideration as sources of inspiration. Therefore, and because our brief summaries are no full substitute for the originals, we strongly encourage reading these seven classic papers, whether this means revisiting them after a long time or discovering them for the first time. The effort will be rewarded.

# 2

# THE PHYSICAL TEMPLATE

## GEOMORPHIC AND HYDROLOGIC UNDERPINNINGS OF STREAM ECOSYSTEMS

Geoffrey C. Poole, Ellen E. Wohl, and Stephen P. Rice

The physical manifestations of streams are diverse. Streamflow can be permanent or ephemeral in time, continuous or intermittent along a stream's course, and well confined by canyons or unconstrained by floodplains. What defines a "stream" is fundamentally geomorphic—a course created by channelized water flow and, implicitly, associated sediment transport. Without a channel, there is no stream.

The geomorphic and hydrologic underpinnings of streams, then, provide a starting point for inquiry into the nature of stream ecosystems. Research into valley, floodplain, and channel form has provided insight into the nature of physical processes that govern the formation and evolution of stream channels (**Ward 1989** [chapter 3]). Thus, physical scientists have contributed markedly to understanding the importance of interactions among physical and biological processes, which are a hallmark of stream ecosystems.

Climate, weather, and catchment geology influence the rate and nature of water and sediment delivery to and through a stream network, with implications for the geomorphic character (e.g., slope, cross-sectional structure, surface sediment) of the valleys, floodplains, and channels formed by streams and the sediments that make up riverbeds. For instance, sediment-choked streams at the toe of a glacier exhibit a braided-channel form with gravelly substrates. Coastal-plain rivers meander over sandy beds. Mountain floodplain streams are "anabranched," having prominent side channels that wander around stable, vegetated islands and more labile bars of sediment. In turn, geomorphic character influences rates and patterns of erosion, sediment transport, and deposition, especially by influencing the character and density of near-stream vegetation that binds sediment and, in forested settings, provides instream wood that can redirect flows. Thus, as demonstrated by the papers highlighted in this chapter, the "flow regime" (the typical seasonal to decadal variation in a stream's discharge), associated erosive forces, the resulting mosaics of geomorphic structures across spatiotemporal scales, and associated patterns of water and sediment movement are important considerations for understanding the structure and dynamics of the physical template on which stream ecosystems self-organize.

## HIGHLIGHTED PAPERS

Wolman, M. G., and J. P. Miller. 1960. "Magnitude and Frequency of Forces in Geomorphic Processes." *Journal of Geology* 68:54–74.

The erosion and deposition induced by an extreme flood are so visually compelling that river scientists once assumed that these unusual flows transported most sediment and shaped river corridors. Wolman and Miller systematically examined this assumption using measurements of streamflow and suspended sediment transport from multiple catchments. They compared the magnitudes and frequencies of different streamflows with associated sediment transport, developing a bivariate plot with three curves—sediment transport rate, flow occurrence frequency, and the product of frequency and rate—plotted against the associated stress applied to the streambed. Their analyses indicated that high flow magnitudes occurring once or twice each year were responsible for most of the total sediment transport. Wolman and Miller then equated those flow magnitudes with morphological "bankfull" conditions, concluding that the channel shapes and dimensions of meandering rivers are associated with, and controlled by, approximately bankfull flows.

This generalization came with caveats. Wolman and Miller noted that "as the variability of the flow increases . . . , a larger percentage of the total load is carried by less frequent flows," and that "closer observation of many geomorphic processes is required before the relative importance of . . . events of differing magnitude and frequency in the formation of given features . . . can be adequately evaluated" (p. 54). Despite these caveats, the concept of "the primacy of bankfull flows" has been applied rather indiscriminately, particularly with river restorations involving channel reconfiguration. In rivers shaped by less frequent but higher-magnitude flows, well-documented channel failures have occurred when the "restored" channel was completely reworked by subsequent floods (e.g., Kondolf et al. 2001).

Numerous subsequent studies documented the role of infrequent floods in shaping channel morphology and transporting sediment in channels with greater erosional resistance (e.g., bedrock channels, gravel-bed channels; Milan 2012; Baynes et al. 2015) and in climates with substantial hydrologic variability (e.g., drylands and the seasonal tropics; Kale et al. 1994). The role of diverse flow magnitudes now permeates thinking regarding "environmental flows" that sustain biological functions, such as organismal movement and reproduction, and maintain physical channel characteristics, such as substrate grain-size distribution, channel-floodplain hydrologic connectivity, and channel planform configuration (Acreman and Ferguson 2010; Meitzen et al. 2013; Hayes et al. 2018).

Hynes, H. B. N. 1975. "The Stream and Its Valley." *Internationale Vereinigung für theoretische und angewandte Limnologie: Verhandlungen* 19:1–15.

Hynes's oft-quoted aphorism "In every respect the valley rules the stream" (Hynes 1975, p. 12) neatly summarizes his paper's central argument: stream ecologists must look beyond the channel at a stream's valley and catchment to understand the stream's ecological processes and resident taxa. Hynes rejects the caricature of a stream as "a gutter, fed by runoff from a landscape paved like a parking lot" (p. 13) and offers a radically different perspective in which the water's chemistry and organic content, and therefore its ecology, are intimately dependent on valley setting.

The 1975 paper draws together innovative contemporary studies (e.g., Johnson et al. 1969; Egglishaw 1972; Cummins et al. 1973; **Fisher and Likens 1973** [chapter 7]) to demonstrate how "upslope" (not simply "upstream") lithology, soils, hydrology, and vegetation define stream bioenergetics, water sources, ion content, and the origin and fate of organic matter. In a few short but lucid pages, he lays out the importance of valley context for fundamental processes, including the generation of proteins from litter decay and the chemical maturation of soil (and therefore stream) water. He draws reasoned inferences from field and laboratory work and is neither evangelical nor prescriptive. The paper's measured tone and careful construction yield an accessible and engaging synthesis, presenting numerous testable ideas.

Indeed, almost fifty years post-publication, the paper still stimulates investigations of its key ideas (Sosnovsky et al. 2022) and engages freshwater scientists in hypothesis testing (Mouron et al. 2022). The work is cited widely and frequently in interdisciplinary contexts, such as understanding catchment (Likens 1984; Stanford et al. 2017) and land-use influences on stream ecology (Allan 2004; Maloney et al. 2020); riverscape approaches (Ward 1998; Poole 2002; Torgersen et al. 2022); effective river restoration and management (Lake et al. 2007; Hawley 2018); (dis)continuity and connectivity of matter and energy in river corridors (Amoros and Roux 1988; Gomi et al. 2003; Harvey and Gooseff 2015); and calls to action aimed at protecting freshwater biodiversity (Dudgeon et al. 2006; Cooke et al. 2022).

Hynes's plea to consider valley and catchment context encouraged consideration of landscape-scale influences on stream ecology more broadly, including geomorphological processes that define bed sediments, sediment fluxes, and channel morphodynamics. His paper can therefore be seen as vital for the growth of a modern stream ecology that reaches across ecology, hydrology, and geomorphology. Petts and Gurnell (2005, p. 34) suggest that the work "provided the catalyst for recognizing fluvial geomorphology as the basis of freshwater ecology." While Hynes's paper is not the only such catalyst, it has been uniquely instrumental in bringing together aspects of river science, with many fruitful outcomes.

Frissell, C. A., W. J. Liss, C. E. Warren, and M. D. Hurley. 1986. "A Hierarchical Framework for Stream Habitat Classification: Viewing Streams in a Watershed Context." *Environmental Management* 10:199–214.

Frissell, Liss, Warren, and Hurley's 1986 paper is among the most quietly influential papers in the field of stream ecology. The authors summarize the paper's aim as follows: "If . . . biological patterns in streams are largely adjusted to and controlled by physical patterns, the problem becomes one of understanding these physical patterns across time and space." Their framework is deceptively simple, positing (1) a finite number and type of observable, repeating stream geomorphic units; (2) a specific class of unit at each spatial scale; (3) hierarchically nested units forming habitat mosaics; and (4) a distinct time scale and magnitude of disturbance necessary to impart geomorphic change at any spatial scale. Standing alone, each tenet could be considered almost self-evident. Yet together, the tenets form a framework for organizing and considering relationships among disparate scientific principles describing streams.

Frissell and his colleagues describe each hierarchical geomorphic unit as a "system" unto itself, emphasizing how processes operating within a given unit provide context for finer-scale units (figure 2.1). Because their work was operational, it provided a basis for integrating competing contemporary concepts within the scientific literature, such as the river continuum concept (**Vannote et al. 1980** [chapter 3]), the flood pulse concept (**Junk et al. 1989** [chapter 9]), and the serial discontinuity concept (Ward and Stanford 1995a). By applying a patchy systems hierarchy to river ecosystems, they integrated these disparate concepts within holistic stream synthesis efforts (Townsend 1996; Thorp et al. 2006).

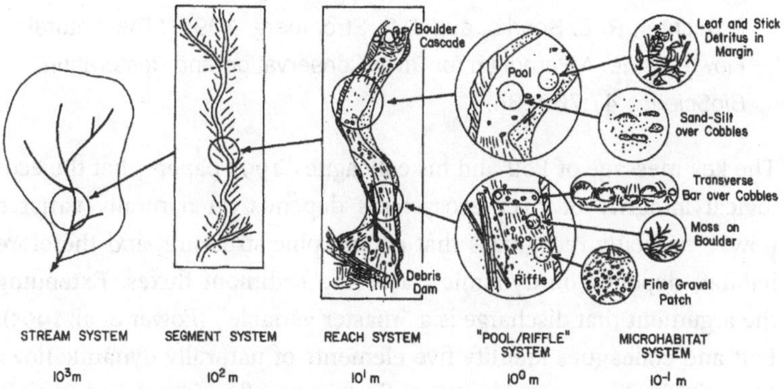


Fig. 2.1 Hierarchical organization of a stream system and its habitat subsystems. Approximate linear spatial scale, appropriate to second- or third-order mountain stream, is indicated. From Frissell, C. A., W. J. Liss, C. E. Warren, and M. D. Hurley. 1986. "A Hierarchical Framework for Stream Habitat Classification: Viewing Streams in a Watershed Context." *Environmental Management* 10:199–214. Reprinted courtesy of Springer Nature.

Frissell and colleagues' work was integral to the fusion of landscape ecology principles with stream ecology (e.g., Ward 1998; Fausch et al. 2002; Poole 2002; Wiens 2002; Allan 2004). Since the paper's publication, its conceptual framework has underpinned many syntheses in the stream ecology literature, including **Ward's (1989** [chapter 3]) "four-dimensional nature" synthesis, **Montgomery's (1999**) "process domains" concept, and **Lake's (2000** [chapter 3]) "disturbance, patchiness, and diversity" treatise, each of which is presented in this book as a foundational paper in its own right.

Frissell and his colleagues also provided principles used in subsequent papers to organize observational data in various subdisciplines, including riparian ecology (**Gregory et al. 1991** [chapter 9]; Naiman et al. 2010), stream community ecology (**Townsend and Hildrew 1994** [chapter 3]; Poff 1997), and stream food webs (Woodward and Hildrew 2002; Baxter et al. 2005). These subsequent contributions are among the most influential in their respective subdisciplines. In the end, the paper's conclusion seems at once prophetic and humble: "This framework is presented as a tool that can guide researchers and managers in conceiving and executing studies, perhaps affording new ways of dealing with old problems."

Poff, N. L., J. D. Allan, M. B. Bain, J. R. Karr, K. L. Prestegaard, B. D. Richter, R. E. Sparks, and J. C. Stromberg. 1997. "The Natural Flow Regime: A Paradigm for River Conservation and Restoration." *BioScience* 47:769–84.

The key message of Poff and his colleagues' 1997 paper—that the ecological integrity of river ecosystems depends on naturally variable flows—explicitly recognizes that geomorphic structure, and therefore habitat, depends on dynamic water and sediment fluxes. Extending the argument that discharge is a "master variable" (Power et al. 1995), Poff and colleagues identify five elements of naturally dynamic flows (magnitude, frequency, duration, timing, and flashiness) and explain their structural and ecological significance. For example, high flows drive floodplain connectivity (**Junk et al. 1989** [chapter 9]), and flashiness can protect naturally adapted native species (Meffe 1984). The paper discusses regional variation (figure 2.2) and changes in flow regimes associated with human activities (e.g., damming, urbanization, land drainage, and channelization) and documents observed ecological responses, including seasonal flow homogenization and increased flashiness caused by hydropeaking. The unequivocal conclusion is that unnatural flow regimes degrade river ecosystems, disproportionately affecting native species that have coevolved with naturally variable flows (e.g., Meffe and Minckley 1987).

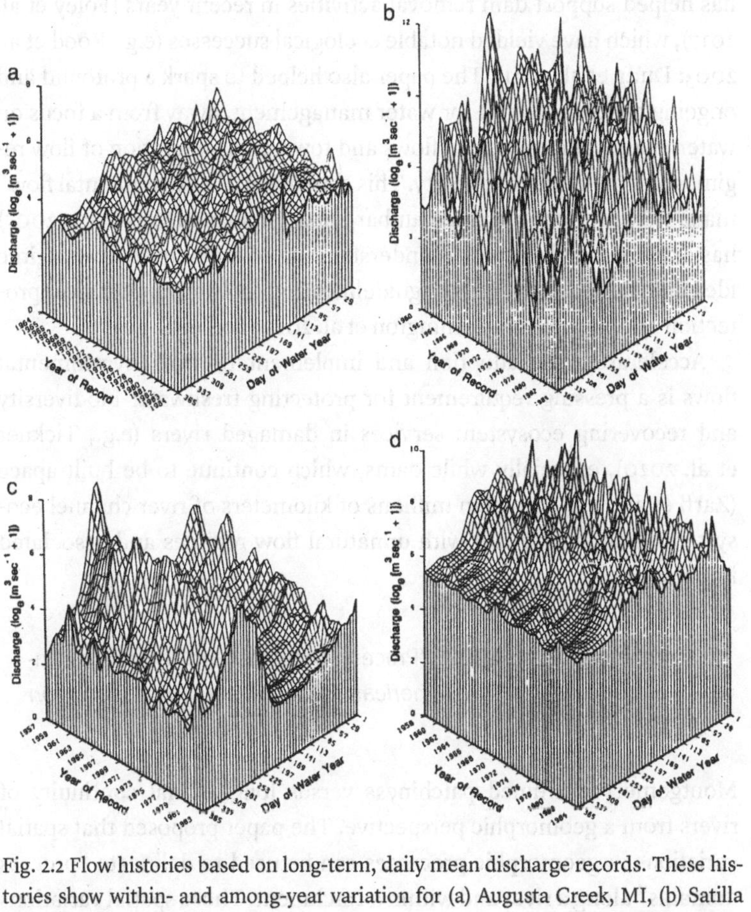


Fig. 2.2 Flow histories based on long-term, daily mean discharge records. These histories show within- and among-year variation for (a) Augusta Creek, MI, (b) Satilla River, GA, (c) upper Colorado River, CO, and (d) South Fork of the McKenzie River, OR. Each water year begins on October 1 and ends on September 30. Adapted from Poff and Ward 1990. From Poff, N. L., J. D. Allan, M. B. Bain, J. R. Karr, K. L. Prestegaard, B. D. Richter, R. E. Sparks, and J. C. Stromberg. 1997. "The Natural Flow Regime: A Paradigm for River Conservation and Restoration." *BioScience* 47:769–84. By permission of Oxford University Press.

Despite their rather bleak synthesis, the authors were hopeful that the twenty-first century "holds promise as an era for renegotiating human relationships with rivers" (p. 780), arguing that re-establishing natural flow regimes, or at least mimicking parts of them, should be a priority for river management. By emphasizing the significance of free-flowing rivers, the paper helped highlight the ecological and geomorphological impacts of dams (Petts 1984; Ward and Stanford 1995b) and has helped support dam removal activities in recent years (Foley et al. 2017), which have yielded notable ecological successes (e.g., Rood et al. 2003; Duda et al. 2021). The paper also helped to spark a profound and ongoing conceptual shift for water management, away from a focus on water quality and minimum flows and toward consideration of flow regimes and dynamic complexity. This shift toward "environmental flow" management (Acreman and Dunbar 2004; Brisbane Declaration 2007) has extended and deepened understanding of the natural flow regime idea and has led to actionable guidelines and tools to support river protection and restoration (Arthington et al. 2010; Poff et al. 2017).

Accelerating the adoption and implementation of environmental flows is a pressing requirement for protecting freshwater biodiversity and recovering ecosystem services in damaged rivers (e.g., Tickner et al. 2020), especially while dams, which continue to be built apace (Zarfl et al. 2015), threaten millions of kilometers of river channel ecosystem (Grill et al. 2019) with unnatural flow regimes and associated implications.

Montgomery, D. R. 1999. "Process Domains and the River Continuum." *Journal of the American Water Resources Association* 35:397–410.

Montgomery examined patchiness versus longitudinal continuity of rivers from a geomorphic perspective. The paper proposed that spatial variations in geomorphic processes can be used to delineate "process domains" along a river and within a catchment. These spatial variations govern temporal patterns of disturbance, which in turn influence ecosystem structure and dynamics.

Montgomery related his geomorphic conceptualizations to the river continuum concept (RCC; **Vannote et al. 1980** [chapter 3]) and hierarchical habitat patch mosaics (**Frissell et al. 1986**), as well as to geomorphic channel classifications designed to reflect underlying processes (Montgomery and Buffington 1997). He described the RCC and patch dynamics as "end-member concepts for the influence of geomorphic processes on ecosystems at the scale of valley segments and channel reaches" (p. 401), noting that neither conceptualization addressed the underlying spatial structure of geomorphic processes. Montgomery proposed process domains as providing both a context for applying patch dynamics across entire catchments and a complementary alternative to the RCC that integrates temporal variability in disturbance processes and the effects of upland processes on river corridors.

The concept of process domains has subsequently been applied to diverse aspects of river form and process, including sediment dynamics in bedrock canyons (Wohl 2010); how riparian communities reflect hydroclimatic disturbance regime and valley morphology (Polvi et al. 2011; Su et al. 2019); the influence of glaciation on Holocene river form and network-scale variations in channel form and process (Brardinoni and Hassan 2006; Livers and Wohl 2015); the correlations between valley lateral confinement and river ecosystem characteristics (Bellmore and Baxter 2014); $CO_2$ evasion from rivers (Rocher-Ros et al. 2019); and large wood dynamics on floodplains (Wohl 2020).

The conceptualization of a river or a catchment as being composed of spatially discrete process domains that can shift their characteristics and locations over time (e.g., in response to changes in climate or tectonics) is related to other geomorphic conceptualizations, such as the differentiation among primarily production, transfer, and depositional zones of a catchment (Schumm 1977), the river styles framework of Brierley and Fryirs (2000), and the more ecologically focused "functional process zones" of Thorp et al. (2006).

Kasahara, T., and S. M. Wondzell. 2003. "Geomorphic Controls on Hyporheic Exchange Flow in Mountain Streams." *Water Resources Research* 39:1005.

"Hyporheic exchange" is the continuous, bidirectional exchange of water between a river channel and underlying sediments. The hyporheic zone is the volume of streambed and floodplain aquifer sediments that contains water derived from the stream channel. The hyporheic zone provides important subsurface, channel-adjacent habitats in streams, and water emerging from the hyporheic zone can have a marked influence on channel temperature, dissolved chemical species, and in-channel habitat.

Orghidan's 1955 paper (English translation published in 2010) is commonly considered the original description of the hyporheic zone, including the various physical, chemical, and biological characteristics of hyporheic habitats. A comprehensive understanding of the geomorphic drivers of hyporheic hydrology would follow decades later, beginning with Toth's (1963) paper describing how variation in surface geography creates localized flow paths that are distinct from regional flows. Harvey and Bencala (1993) applied these concepts to hyporheic exchange in a comprehensive set of empirical observations backed by simulation-model results, showing how hyporheic exchange is driven by the water flow atop geomorphic features. Subsequent work has shown that stream structures observable at different scales contribute to hyporheic exchange via various mechanisms. For instance, surface grain size (Packman and Salehin 2003), pool-riffle sequences (Tonina and Buffington 2011), meander bends (Cardenas et al. 2004), and valley constrictions (Malard et al. 2002) all operate to influence rates and locations of hyporheic exchange.

Clearly, then, no single paper represents the foundations of hyporheic hydrologic research. We highlight Kasahara and Wondzell's paper as an excellent example of research that integrates the three main methods through which knowledge of hyporheic hydrology has been advanced: theory, empirical observations, and simulation modeling. The paper's research design includes development of a detailed three-dimensional simulation model, calibrated against extensive field observations, describing flow fields in the hyporheic zone of four stream reaches. The paper is rooted in Toth's (1963) theory and uses in silico experiments in which different geomorphic features are removed from the simulation model to assess their influence on hyporheic exchange. The results show that pool-step sequences drive hyporheic exchange in small (second-order) mountain streams, whereas a broader set of morphological features, together with the degree of channel constraint, combines to shape patterns of exchange in larger (fifth-order) streams. Although other papers listed above first described geomorphic controls on hyporheic exchange, few exhibit the comprehensive approach taken by Kasahara and Wondzell, nor define more clearly the breadth of processes that link stream forms to hydrologic function.

## LOOKING FORWARD

The foundational papers reviewed in this chapter highlight how a stream's physical template is driven by linked hydrologic and geomorphic processes that extend across scales from channel microhabitats to hillslopes and so integrate the domains of surface and subsurface environments. The primacy of the physical template as a constraint on community structure and ecological processes in streams can hardly be overstated (Poff, Bledsoe et al. 2006; Poole 2010). At the same time, any focus on the physical template in isolation provides a lens too myopic for effective stream management.

Stream restoration, as a management practice, is often implemented as a means of imposing desired physical structures. Direct reshaping of the physical template has, in many ways, become an end unto itself (Simon et al. 2007), ignoring the fact that stream ecosystems arise from interactions among physical and biological processes (**Ward 1989**; Rice et al. 2001; Pollock et al. 2014; Wohl 2017; see chapter 9). Most restoration projects focus exclusively on modifying flows and engineering a desired channel and floodplain morphology (Bernhardt, Palmer et al. 2005), with scant consideration of physical processes that shape and reshape streams, biotic controls on geomorphology (Murray et al. 2008; Rice 2021), or ecological drivers of population dynamics such as food sources, recruitment, and mortality rates (Wohl et al. 2005; Beechie et al. 2010). From this perspective, sustainable management and restoration of stream ecosystems requires the development of strategies that encourage the subsequent formation—rather than immediate imposition—of desirable physical habitat structures, within a broader appreciation of the breadth of ecological constraints that affect restoration schemes (Wohl et al. 2005; Paola et al. 2006; Palmer et al. 2014). Alternative approaches to river restoration, including process-based restoration (Beechie et al. 2010), biomic river restoration (Johnson et al. 2020), and restoration of conditions facilitating formation and maintenance of river-wetland corridors (Flitcroft et al. 2022), are indicators of a promising shift toward integrative and dynamic stream management.

# 3

# STREAMS AS UNIQUE HABITATS

Debra S. Finn and David M. Merritt

The geomorphic and hydrologic underpinnings of the physical setting in streams result in unique habitat characteristics for the organisms that dwell in them. These characteristics include not only the physical challenges that come with life in moving water, but also substantial heterogeneity in space and time, as well as in the components of spatial habitat structure that shift across scales (**Frissell et al. 1986** [chapter 2]). Early recognition of these habitat features by stream ecologists coincided with both the development of landscape ecology (Forman and Godron 1981, 1986) and a general shift toward nonequilibrium thinking in ecology (Sousa 1984; Pickett and White 1985). Although this chapter emphasizes direct organismal responses to the stream environment, the distinctive physical habitat of streams is a thread that runs through all the chapters of this book. Unique and dynamic habitat requires unique traits or "tactics" (*sensu* Southwood 1977) of organisms. Hence, we also highlight the relevance of a functional approach that seeks to link organismal traits to the environment. First, we briefly discuss three unique aspects of running-water habitats, the recognition of which facilitated the maturation of stream ecology. Together, they form a conceptual basis for the papers we highlight below.

***Flow-associated disturbance is ubiquitous in streams.*** Running water is the dominant force shaping stream ecosystems and organisms. Occupying space within the advective flow of a liquid medium presents its own challenges (Hart and Finelli 1999; Statzner et al. 1988), which are exacerbated by temporal variation in flow magnitude. High-flow events cause substantial changes to the physical habitat, dislodging, burying, damaging, and killing organisms. Low-flow periods bring other types of disturbance, including hypoxia and the concentration of organisms and waste products within reduced habitat volumes. Hence, streams are uniquely dynamic and characterized by multiple disturbance gradients along a range of magnitudes and frequencies. Components of a stream's natural flow regime are thus strong selective agents for organismal traits associated with resistance and resilience to disturbance (Lytle and Poff 2004; **Poff et al. 1997** [chapter 2]). Examples include behavioral strategies for temporary relocation (Lancaster and Hildrew 1993) and life-history adaptations that synchronize events such as adult insect emergence (Lytle 2002) and seed release by riparian plants (Karrenberg et al. 2002) with seasonal floods.

***Stream habitat is connected along three spatial dimensions.*** Stream ecosystems are characterized by connectivity in lateral, longitudinal, and vertical dimensions. Among the earliest conceptual developments in stream ecology was the recognition of the vital role of allochthonous energy inputs from adjacent terrestrial ecosystems (see chapter 1). Material exchange in this lateral dimension was later understood to occur in the opposite direction as well (e.g., emergent aquatic insects subsidizing terrestrial food webs), while lateral connectivity between the stream channel and a periodically inundated floodplain was recognized as providing mobile organisms access to both seasonally productive habitat and refuge during flood disturbance (see chapter 9). Connectivity in the upstream-downstream (longitudinal) and surface-subsurface (vertical) dimensions plays similar roles in transport or transformation of basal resources and in the provision of bidirectional movement pathways for organisms. Importantly, connectivity in all three spatial dimensions allows flexibility in the capacity of organisms to respond to dynamism in the temporal dimension (**Ward 1989**).

***Streams are hierarchically branching networks formed from merged linear segments.*** Hydrologic connectivity along linear pathways from headwaters to large rivers inspired early studies of environmental gradients and longitudinal turnover of species composition, particularly in fishes (Illies and Botosaneanu 1963; Sheldon 1968).

Gradients are certainly not unique to streams, having inspired some of the earliest work in terrestrial ecology (Humboldt and Bonpland 1807), but stream gradients have unique aspects, including predictable changes in habitat size and shape. The river continuum concept (RCC; **Vannote et al. 1980**) asserted that these predictable changes drive ecosystem functional responses. But streams are not just lines. Streams branch like the vascular system of a complex organism, with tiny headwaters spidering across landscapes, coalescing into increasingly large but less numerous vessels. This hierarchically nested dendritic structure has important consequences for ecological function (Fisher 1997; Rodriguez-Iturbe and Rinaldo 1997). Headwaters in particular, given their intimate connection with the terrestrial setting and spatial isolation from one another as distal tips of networks, can vary radically from one to the next and strongly influence processes and biodiversity at the scale of entire river networks (**Meyer and Wallace 2001**).

The combination of these unique physical habitat features results in a view of streams as complex and connected "riverscapes" that are heterogeneous in space and time (Fausch et al. 2002; Allan 2004). Put another way, Fisher (1997) stated that what we perceive of streams is "the surface manifestation of a larger, integrated, planar system that represents an underlying and unifying fabric of the landscape." Although disturbance prone, lotic systems also contain a substantially greater global diversity of organisms than expected given the minimal surface area they occupy (Strayer and Dudgeon 2010). In this chapter, we highlight foundational papers that underscore various aspects of organism response and functional diversity associated with the complex and dynamic physical habitat that characterizes streams and rivers. Collectively, these papers helped develop the riverscape perspective in stream ecology. And because it is increasingly clear that this disproportionate diversity in streams can persist only if habitat heterogeneity and pathways for connectivity are maintained, we conclude the chapter with a short section emphasizing pressing applied issues in stream ecology. We suggest that conservation of biological diversity under rapid change will require a holistic approach to maintaining the unique patterns of connectivity that streams require to function naturally.

## HIGHLIGHTED PAPERS

Vannote, R. L., G. W. Minshall, K. W. Cummins, J. R. Sedell, and C. E. Cushing. 1980. "The River Continuum Concept." *Canadian Journal of Fisheries and Aquatic Sciences* 37:130–37.

"The River Continuum Concept" was a pivotal publication in stream ecology. Vannote and his colleagues helped shift our field from a descriptive to a more predictive science by proposing an ecosystem perspective for understanding biological turnover along the physical habitat gradient from headwaters to large rivers. The authors were inspired by progress in several arenas, including the tracing of energy flow through ecosystems (e.g., Odum 1957), catchment-scale ecosystem studies (Bormann and Likens 1967; see chapter 10), fluvial geomorphology (Leopold et al. 1964), foundational studies of upstream-downstream turnover in species assemblages (e.g., Illies and Botosaneanu 1963), and an appreciation of the crucial contribution of basal food resources from terrestrial sources to streams (e.g., **Minshall 1967** [chapter 5]). Vannote and coauthors leaned on their diversity of expertise—from organismal biology to ecosystem ecology to geomorphology—to roll these varied perspectives into a single, unified framework for anticipating ecological responses along longitudinal gradients. Despite the RCC's wide-ranging influence, their manuscript was rejected by multiple journals before an enthusiastic acceptance at the *Canadian Journal of Fisheries and Aquatic Sciences*. The paper caused a stir in the developing field of stream ecology, inciting heated responses immediately following its publication and pushing the development of new perspectives that challenged, complemented, and built upon it.

The RCC characterized stream ecosystems as existing in a state of dynamic equilibrium in response to the broadly predictable physical habitat gradient. The idea meshed well (perhaps "beautifully"; McAllister 1996) with a corresponding concept in fluvial geomorphology in which the channel at a particular location along the gradient attains a physical equilibrium of morphological and hydraulic conditions (Langbein and Leopold 1964). Vannote and colleagues proposed that the biotic assemblage at any location along the longitudinal gradient reflects this equilibrial channel morphology. To illustrate predicted ecological

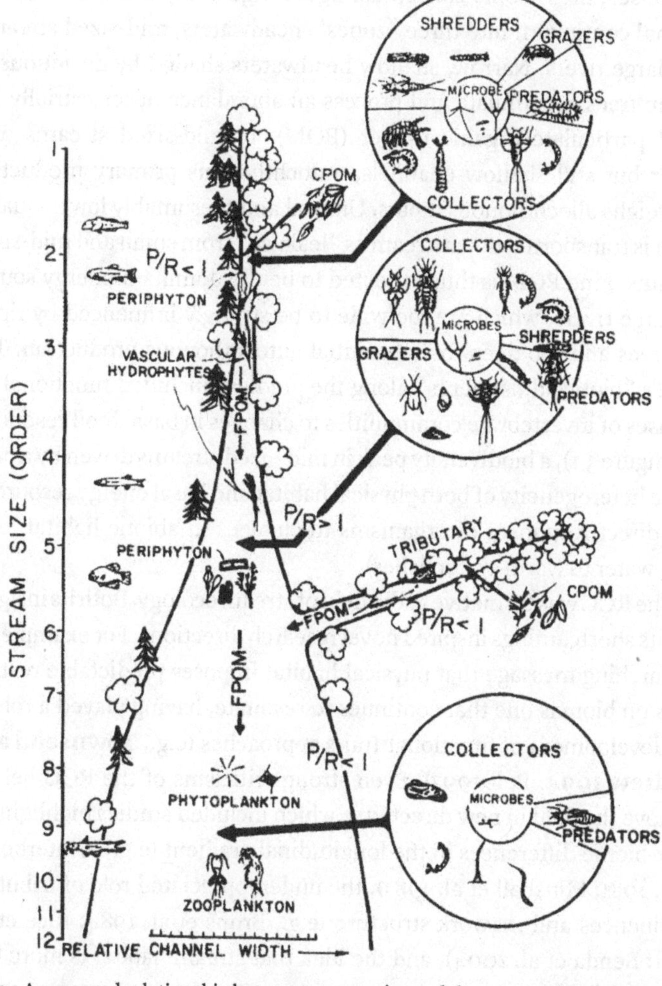


Fig. 3.1 A proposed relationship between stream size and the progressive shift in structural and functional attributes of lotic communities. See text for fuller explanation. From Vannote, R. L., G. W. Minshall, K. W. Cummins, J. R. Sedell, and C. E. Cushing. 1980. "The River Continuum Concept." *Canadian Journal of Fisheries and Aquatic Sciences* 37:130–37.

responses, their iconic conceptual figure (figure 3.1) divided the longitudinal continuum into three "zones": headwaters, mid-sized streams, and large rivers. Narrow, shallow headwaters shaded by deciduous riparian trees accumulate and process an abundance of terrestrially derived particulate organic matter (POM). In mid-sized streams with wider but still shallow channels, autochthonous primary production outweighs allochthonous inputs. Unused and presumably lower-quality POM is transported downstream as "leakage" from small and mid-sized streams. Fine POM is thus expected to be the dominant energy source for large rivers, which are too wide to be strongly influenced by riparian trees and too deep for substantial autochthonous production. The RCC's "biotic adjustments" along the gradient included functional responses of invertebrate communities to changes in basal food resources (see figure 3.1), a biodiversity peak in mid-sized streams driven by a peak in the heterogeneity of both physical habitat and basal energy resources, and direct responses of organisms to changes in abiotic habitat (e.g., cold-water to warm-water fishes).

The RCC was formative to the field of stream ecology. Both its insights and its shortcomings inspired novel research directions. For example, its overarching message that physical habitat imposes predictable restrictions on biota is one that continues to resonate, having played a role in the development of functional-traits approaches (e.g., **Townsend and Hildrew 1994**; Poff 1997). Even strong criticisms of the RCC helped to move the field in new directions, which included studies highlighting inter-biome differences in the longitudinal gradient (e.g., Winterbourn et al. 1981; Minshall et al. 1983), the underappreciated role of tributary confluences and network structure (e.g., Bruns et al. 1984; Rice et al. 2001; Benda et al. 2004), and the idea that stream habitat is more like a "mosaic" than a smooth continuum (e.g., Downes et al. 1993). Importantly, the RCC lacked an emphasis on the roles of flow variability and disturbance (Statzner and Higler 1985), a unique property of streams that would gain strong research traction in the decade following the paper's publication (Barnes and Minshall 1983; Fisher et al. 1982; Poff and Ward 1989). Theory that stimulates thinking, catalyzes research in multiple directions, and ultimately advances a field has value well beyond its individual strengths and weaknesses. Vannote and his colleagues unquestionably advanced our thinking about stream ecosystem structure and function.

Ward, J. V. 1989. "The Four-Dimensional Nature of Lotic Ecosystems." *Journal of the North American Benthological Society* 8:2–8.

The 1970s and 1980s were decades of substantial progress in the fields of freshwater and riparian ecology, including the recognition that streams differ from lentic systems by being connected networks spanning spatial gradients of terrestrial influence (**Hynes 1975** [chapter 2]; **Vannote et al. 1980**). J. V. Ward was strongly influenced by the developing field of landscape ecology (Forman and Godron 1986), with its emphasis on spatial heterogeneity and connectivity, but he had also been accumulating evidence in streams and rivers that *dis*connection, or "discontinuities" (e.g., Ward and Stanford 1983), resulting from human activities deserved more scientific attention. Ward, in the spirit of Leonardo da Vinci's adage "Everything connects to everything else," proposed a holistic new framework that explicitly defined streams as open and temporally dynamic systems. His 1989 paper introduced a four-dimensional view of stream ecosystems that included the three spatial dimensions of connectivity unique to streams (longitudinal, lateral, and vertical) in addition to time, the crucial fourth dimension, whose inclusion acknowledged the importance of temporal flow variability.

To make the case for the ecological relevance of the four-dimensional framework, Ward reviewed evidence from the literature. The RCC had emphasized the longitudinal dimension, which is important for invertebrate drift (**Waters 1962** [chapter 4]), hydrochory (dispersal of plant reproductive propagules; Nilsson et al. 2010), and migrations of fish and other mobile consumers (Kareiva et al. 2000). The lateral dimension had been recognized as important in the transfer of energy and organisms to streams from riparian zones, and vice versa (e.g., Jackson and Fisher 1986; see chapters 6 and 9). The vertical dimension connecting stream surface and subsurface zones had been vastly underappreciated, but Ward explicitly pointed to the hyporheic zone as an important ecotone between surface-water and groundwater ecosystems (see chapter 2) and as an "interstitial highway" of connectivity for benthic fauna (e.g., Williams and Hynes 1974; Stanford and Ward 1993). Ward also recognized that time—the fourth dimension—interacted with the spatial hierarchy of physical habitat patches (*sensu* **Frissell et al. 1986** [chapter 2]), components of which vary in temporal persistence. Ward also argued that the strength of connectivity in any single dimension (e.g., laterally from channel to floodplain) likely varies as a function of position within other dimensions (e.g., longitudinally from headwaters to large rivers).

In formalizing the four-dimensional nature of streams, Ward set the stage for the riverscape perspective, which he characterized in a later paper (Ward 1998) as "a holistic geomorphic perspective of the extensive interconnected series of biotopes and environmental gradients that, with their biotic communities, constitute fluvial systems." This holistic, connected, and dynamic perspective inspired countless research directions. Indeed, each of the remaining highlighted papers in this chapter (and many in other chapters) represents a quintessential example of the importance of Ward's work to the development of stream ecology.

Schlosser, I. J. 1991. "Stream Fish Ecology: A Landscape Perspective." *BioScience* 41:704–12.

Schlosser's paper represents an early and influential example of applying a riverscape perspective, inspired by the theoretical work of **Ward (1989)** and **Frissell et al. (1986** [chapter 2]), to management issues. It is also distinguished from most papers highlighted in this book by its focus on fish and fisheries. Early in his career, Schlosser was inspired by work quantifying links between fish species diversity and habitat complexity at the reach scale (Gorman and Karr 1978). As the general appreciation of scale and connectivity increased throughout the 1980s, Schlosser and others working on fish ecology began conceptualizing the importance of unique aspects of stream habitat to mobile consumers with complex life cycles (e.g., Bain et al. 1988; **Junk et al. 1989** [chapter 9]). In his 1991 paper, Schlosser forged a new conceptual understanding of the role of spatial and temporal habitat heterogeneity in the production and diversity of stream fishes.

Schlosser built the case that both spatial heterogeneity and connectivity at multiple scales are essential for stream fishes throughout their life cycles. He elegantly incorporated the notion that flow dynamism enhances habitat complexity and influences patterns of spatial connectivity among habitat patches. Central to Schlosser's model is the idea that fish species typically require spatially distinct habitats for different life stages, such as spawning, juvenile rearing, feeding, and refuge from unfavorable conditions, and that each of these habitats must be linked by fish movement. Schlosser called on his own data (Schlosser 1982, 1987) and other empirical evidence (e.g., Bisson et al. 1982; Welcomme 1985; Power et al. 1989) to demonstrate that aspects of habitat heterogeneity govern variation in fish populations and communities at spatial scales ranging from different depths within a single pool to juvenile versus adult habitats thousands of kilometers apart. He also emphasized that connectivity among key habitats is required at least periodically.

An important aspect of Schlosser's paper was the application of major stream ecological concepts to problems associated with human activities and their effects on habitat heterogeneity and connectivity in streams. Schlosser extended the ideas in Aldo Leopold's (1949) land ethic to generate a new "riverscape ethic" for stream and river fishes, and his paper motivated further work to understand how connectivity and heterogeneity promote resilience to climate change (e.g., Armstrong and Schindler 2013; Uno and Power 2015). It also inspired continued application of ecological theory to fish management (e.g., Allan 2004) and has even been adopted by environmental advocacy groups (e.g., Noss and Cooperrider 1994).

Townsend, C. R., and A. G. Hildrew. 1994. "Species Traits in Relation to a Habitat Templet for River Systems." *Freshwater Biology* 31:265–75.

Colin Townsend and Alan Hildrew were coeditors of the journal *Freshwater Biology* for thirty-three years, beginning in the early 1980s (Hildrew and Townsend 2007). Those years saw massive growth of the journal that coincided with the maturation of stream ecology and the field's growing appreciation of the ecological relevance of unique habitat features. Their 1994 paper was a focal piece in a special issue of *Freshwater Biology* addressing the potential to apply general theory and develop statistical approaches to glean ecological insight from large datasets, with specific emphasis on a seventeen-year dataset from the Rhône River network in France (Statzner et al. 1994). Although Townsend and Hildrew had not been involved in generating the data, the Rhône River group challenged them to apply theory to generate testable predictions of invertebrate distributions based on functional traits.

To set the stage, Townsend and Hildrew characterized key aspects of a "river habitat templet" (i.e., a range of environmental conditions influencing adaptive strategies of organisms; "template" in US

English), then proposed functional "tactics" expected of organisms occupying various positions on the template (following Southwood 1977). Recognizing the important role of flow variability and spatial habitat complexity in streams, the authors suggested two main axes for describing the template: temporal heterogeneity and spatial heterogeneity. They then emphasized the most relevant of Southwood's proposed categories of functional traits: those associated with strategies to escape disturbance. These included a variety of life-history, morphological, physiological, and behavioral traits putatively associated with both resistance (e.g., morphology such as suckers or a streamlined body shape) and resilience (e.g., short life spans or year-round presence of winged terrestrial forms) to flow-related disturbance. Overall, Townsend and Hildrew proposed that traits conferring either resistance or resilience should be more prominent in more temporally dynamic habitats, and that increased spatial heterogeneity should provide increased opportunities for resistance traits associated with refuge from disturbance.

In their 1994 paper, Townsend and Hildrew applied the developing themes associated with the dynamic riverscape perspective to promote a multidimensional functional-traits approach that quickly influenced the field (Poff 1997; Richards et al. 1997; Usseglio-Polatera et al. 2000; Poff, Olden et al. 2006). They also pointed out several challenges to applying the trait-based template approach that have yet to be fully resolved. These challenges include the potential for alternative suites of traits to respond similarly to a given habitat characteristic. For example, species with long life cycles and sucker-like body parts might respond equally well to frequent floods as species with short life cycles that lack morphological adaptations. They also noted the difficulty of predicting the point along the temporal stability dimension at which biotic interactions might play a greater role than abiotic habitat in determining habitat occupancy. Despite these and other challenges, the trait-based template approach has been extremely influential and has spawned advances in applied subfields such as bioassessment (Bonada et al. 2006) and environmental flow management (Richter et al. 1997). Similar approaches have also been developed for additional taxonomic groups, including riparian vegetation (Karrenberg et al. 2002; Merritt et al. 2010).

Stanley, E. H., S. G. Fisher, and N. B. Grimm. 1997. "Ecosystem Expansion and Contraction in Streams." *BioScience* 47:427–35.

Stanley, Fisher, and Grimm's 1997 paper, a product of Emily Stanley's PhD research, provided the first detailed description of drying patterns and their ecological consequences throughout a well-studied desert stream network (Sycamore Creek, Arizona, USA; see chapter 10). Like other papers in this chapter that converged on the riverscape perspective, this paper was clearly influenced by the field of landscape ecology, with its emphasis on scale, temporal dynamism, and connectivity. Despite a strong focus during the prior decade on the ecological role of flow variability (Poff and Ward 1989) and disturbance (Resh et al. 1988), low-flow and drying events had been largely overlooked in stream ecology. But given the arid climatic setting, high temperatures, and readily observable and rapid ecological responses to changes in wetted area (e.g., algal blooms immediately following rewetting) at Sycamore Creek, extreme fluctuations in aquatic ecosystem size had long been apparent there. These observations laid the groundwork for Stanley's research, and the resulting paper is foundational because expansion and contraction (controlled by temporal flow variability) are now recognized as general features of most streams, regardless of climatic setting.

Data collected seasonally throughout the stream network allowed rigorous evaluation of the spatial and temporal heterogeneity of surface drying at three nested spatial scales ("reach" within "section" within "phase"). At all scales, Stanley and her colleagues provided evidence for a patchy or mosaic-like pattern of drying that depended on streambed substrate, surface-subsurface connectivity, and valley-floor constraints. At the reach scale, runs and riffles lost surface water first, and aquatic organisms aggregated stochastically in disconnected pools. Riffle-dwelling organisms with limited mobility and strict habitat requirements, including aquatic moth larvae that require submerged rocks, were immediately vulnerable to low flows. At the intermediate spatial scale, unconstrained sections with high habitat complexity alternated with less complex, less productive sections constrained within canyons. Mobile consumers migrated to the less productive sections during drying because they retained surface water. At the broadest scale, steep headwaters dried quickly and completely, as did the downstream-most phase of the mainstem due to its deep and rarely saturated alluvial deposits. In contrast, canyon and transitional phases provided aquatic refugia during the driest periods. Overall, the authors' results underscored the importance of heterogeneity and connectivity in stream networks, but also pointed to underappreciated practical issues associated with interpreting ecological responses to drying disturbance. For example, it is difficult to monitor drying patterns over time and space with standard gauges, given the mosaic-like pattern of drying across spatial scales.

Nonetheless, Stanley and colleagues' observations suggested that predictions can be made about patterns of drying in intermittent streams based on hydrology and geomorphology, and that organismal responses based on functional traits, such as mobility, habitat specificity, and desiccation resistance, may also be predictable (see also **Townsend and Hildrew 1994**; Poff 1997). Their work inspired an explosion of research on intermittent rivers and ephemeral streams during the first two decades of the twenty-first century (Dodds et al. 2004; Datry et al. 2014, 2017), which was also motivated by the growing appreciation that naturally drying streams contain diverse assemblages of organisms distinct from those occupying perennial streams. Other predictions in this paper have also been corroborated, including the occurrence of population bottlenecks in smaller and more isolated aquatic habitat fragments during unusually dry periods (Finn et al. 2009; Shama et al. 2011). Each of these ideas continues to be relevant in the context of climate change and water extraction.

Lake, P. S. 2000. "Disturbance, Patchiness, and Diversity in Streams." *Journal of the North American Benthological Society* 19:573–92.

Lake's 2000 paper was a "Perspectives" piece in the *Journal of the North American Benthological Society*, in which he summarized contrasting biological responses to various types of flow disturbance in streams and suggested forward-thinking research directions for anticipating responses to human-caused flow disturbance. The timing and tone of this paper suggest a turning point at which stream ecologists fully embraced stochasticity and nonequilibrium thinking. The paper also helped to settle a two-decade disagreement surrounding terminology associated with flow disturbance (e.g., Resh et al. 1988; Poff 1992). Indeed, Lake began with a well-supported argument that the term "disturbance" is misused when it emphasizes the magnitude of a biological result (e.g., killing or displacement of organisms). Rather, "disturbance" should be reserved for quantifying the abiotic cause of a potential biological response. He then distinguished a "perturbation" as the combination of disturbance and response. For example, a flood of a particular magnitude might be a disturbance associated with relatively minor perturbation if it occurs with predictable seasonal timing (to which organisms are adapted), but the same disturbance likely represents a more severe perturbation if it is unpredictably timed. Clearly, it was essential for stream ecologists to unite around a common terminology for continued scientific progress toward understanding these uniquely disturbance-prone systems.

Lake also distinguished among "pulse" (discrete, short-term), "ramp" (intensifying over time), and "press" (long-term, often human-caused) disturbances, because these differences affect biological responses. For example, because floods are typically pulse and drying typically ramp disturbances, Lake argued that theory related to the biological effects of floods was not useful for predicting responses to drying. He reiterated several points made by **Stanley et al. (1997)** on the importance of spatiotemporal scale and adopting a functional approach for understanding organismal responses to drying. He also pointed out that high-flow disturbances typically increase connectivity at multiple spatial scales, but low-flow disturbances produce gradually increasing "disconnection." Hence, Lake suggested that drying generates a greater perturbation because post-disturbance recolonization is impeded. He further suggested that biological effects of low-flow disturbances might not be comparable across different regions of the globe due to differences in both climate and the geomorphology of river networks. For example, many Australian streams have lengthy dry periods interrupted by severe flood events and appear to offer more diverse strategies for organismal refuge than do North American streams such as Sycamore Creek (Boulton 1989; see chapter 10).

Lake's insights and global perspective endure as climate change and human water consumption amplify the magnitude and frequency of extreme flow events. His contributions have helped foster intercontinental collaboration in anticipation of changing management strategies in rivers with increasing disturbance frequencies (e.g., Palmer et al. 2008), as well as emerging global efforts to develop management strategies specific to intermittent rivers (e.g., Leigh et al. 2016).

> Meyer, J. L., and J. B. Wallace. 2001. "Lost Linkages and Lotic Ecology: Rediscovering Small Streams." In *Ecology: Achievement and Challenge*, edited by M. C. Press, N. J. Huntley, and S. Levin, 295–317. The 41st Symposium of the British Ecological Society. Orlando, FL: Blackwell Science.

Meyer and Wallace's 2001 paper focused our attention on the crucial but underappreciated role of small headwater streams, as well as the conservation challenges these habitats face. By the nature of the hierarchically branching structure of streams, headwaters vastly outnumber larger streams, but they have been largely ignored or undervalued in policy and management. Meyer and Wallace made the striking case that the diverse and influential stream ecology research at the Coweeta Hydrologic Laboratory (in North Carolina, USA, where both worked [see chapter 10]) had been accomplished on streams that "do not exist" because they are too small to appear on most maps (Benstead and Leigh 2012). Their paper was a call to action, both to expand basic ecological research on headwater streams and to build interdisciplinary relationships for addressing policy and management issues associated with headwaters.

Meyer and Wallace set the stage by describing the intimate linkages between small headwater streams and the terrestrial environment (*sensu* **Hynes 1975** [chapter 2]; **Vannote et al. 1980**) and emphasizing the importance of connectivity in multiple dimensions throughout stream networks. This network perspective was novel at the time (Fisher 1997) but critical for communicating the unique value of these numerous smallest and most distal branches. Meyer and Wallace pointed out the role of small streams as both thermal and flow refugia for mobile groups such as fish, and they argued that incremental damage to headwaters drastically reduces habitat complexity at the landscape scale, affecting ecosystem processes and biodiversity in a manner similar to the loss of complex floodplain habitats associated with large rivers (e.g., **Sedell and Froggatt 1984** [chapter 9]). Indeed, habitat heterogeneity among headwaters and their spatial isolation as distant tips of stream networks combine to support high beta (among-site) diversity.

Meyer and Wallace influenced policy associated with small streams and inspired several influential syntheses and opinion pieces on the role of headwaters in both applied and basic ecological contexts (Lowe and Likens 2005; Meyer et al. 2007; Clarke et al. 2008). The network perspective central to this paper also influenced several lines of research associated with the consequences of network spatial structure for biodiversity (e.g., Campbell Grant et al. 2007), population genetics (Hughes et al. 2009; Finn et al. 2011), and the maintenance of species diversity in metacommunities (Brown and Swan 2010; Tonkin et al. 2018). The appreciation that physical habitat can vary substantially among headwaters within networks has also influenced research aimed at informing conservation management under climate change (e.g., Brighenti et al. 2021) and has helped galvanize the push to better understand intermittent streams (Datry et al. 2017). These non-perennial channels not only contain unique components of regional biodiversity but also tend to have greater diversity than similarly sized streams with permanent flow, despite typically "not existing" by virtue of their absence from most maps.

## LOOKING FORWARD

The foundational papers highlighted in this chapter, and the research they inspired, have vastly expanded our mechanistic understanding of organismal responses to the unique habitat components of dynamic riverscapes. We now face the pressing challenge of understanding how stream-dwelling organisms will respond to rapid changes in extreme flows and other habitat characteristics, driven by both climate change and intensifying human water use and development. Conservation of the disproportionate organismal diversity found in lotic ecosystems will require a holistic approach to understanding and managing the spatial and temporal habitat heterogeneity of Anthropocene riverscapes.

Three-dimensional spatial connectivity in stream networks promotes resistance and resilience of stream-dwelling organisms to disturbance, but we know that disconnections and discontinuity result from the most common and widespread water development projects. Dams, for example, have shifted and interrupted patterns of natural connectivity in all riverscape dimensions (Ward and Stanford 1983). But with strategic management, dams can also provide an opportunity to help restore the timing, magnitude, and other components of a more "natural" flow regime (**Poff et al. 1997** [chapter 2]) and reclaim some ecologically important elements of spatial and temporal connectivity. Environmental flows (i.e., "e-flows") emphasize these linkages between flow components and desired characteristics of stream ecosystems (e.g., Poff and Matthews 2013). Even modest alterations in dam operations can produce significant biological responses without substantially impacting the societal benefits associated with water storage in reservoirs (Kennedy et al. 2016). Ongoing challenges in the field of e-flows include planning prescribed (or "designer") flows for suites of species that vary in flow-related life-history traits (Tonkin et al. 2019, 2021) and adjusting flow strategies in regulated rivers that have undergone additional modifications to spatial connectivity, such as levees (Molles et al. 1998). Proposed future hydropower projects in large river networks like the Amazon represent another looming challenge, but there is strong potential to apply network-scale modeling to plan dam locations so as to minimize ecological disconnections (Flecker et al. 2022).

Intermittent rivers and ephemeral streams represent another ongoing area of emphasis because they are widespread and contain unique organisms, and because many once-perennial streams are shifting toward intermittency (e.g., Zipper et al. 2021; Sánchez-Montoya et al. 2023). Ecological understanding of low-flow disturbance has escalated rapidly in the twenty-first century, but many challenges remain. Practical challenges include confronting the "placement bias" of flow-gauging stations worldwide toward large, perennial rivers (Krabbenhoft et al. 2022) and implementing alternative methods for flow monitoring (e.g., remote sensing, repeat photography, and citizen science; Allen et al. 2019). There is also a push toward improved mapping of complete stream networks, including intermittent or ephemeral segments, which might help overcome policy stalemates. Better geospatial tools are also likely to generate improved understanding of the distribution and connectivity of segments within stream networks that are most likely to resist shifts to intermittency and/or warming and, as such, provide refuge habitat (e.g., Brighenti et al. 2021). Empirical work on functional traits of organisms in these systems has strong potential to expand and improve existing trait databases. Although these databases are widely tapped for broad-scale modeling of ecological responses, many gaps remain (Downes and Lancaster 2018; Priest and Finn 2023), likely due in part to a bias toward perennial streams. An accelerating pace of "small data" studies on taxa in intermittent and other underrepresented systems is promising continued improvement of accuracy in linking "tactics" to "template" and tackling challenges associated with shifting connectivity patterns in riverscapes.

Overall, meaningful progress will require a thoughtful combination of old and new approaches, which together hold the greatest promise for continued expansion of our understanding of organism-environment interactions in streams and its application to biodiversity conservation. Recent and ongoing technological advances in areas as wide-ranging as artificial intelligence (Jordan and Mitchell 2015), environmental DNA techniques (eDNA, Deiner et al. 2016), and high-precision streamflow measurement (Oberg and Mueller 2007) hold great promise to enhance our understanding of habitat elements and organismal responses. But back-to-basics natural history studies of stream-dwelling organisms are still necessary to bolster the baseline data needed to leverage emerging approaches. Guided by the classic papers highlighted in this chapter, and the knowledge and developments they have inspired, stream ecologists possess the most advanced and diverse toolbox ever for managing these unique ecosystems and conserving them into the future.

# 4

# SPECIES INTERACTIONS AND COMMUNITY STRUCTURE

Angus R. McIntosh and Jonathan D. Tonkin

Streams offer a dynamic running-water environment and contain organisms adapted to those unique conditions. Species interactions and the development of stream communities are thus strongly influenced by flow and flow-related disturbances, such as floods and droughts. These conditions and influences mean that the role of species interactions, and the processes that drive community assembly, differ considerably from those present in most other environments. For example, the keystone predation that is so prominent in marine rocky intertidal ecosystems (Paine 1966), and the strong competition through size-based feeding efficiency that structures zooplankton in lakes (Brooks and Dodson 1965), have relatively few equivalents in streams. This does not mean that strong interactions do not occur, that there are no keystone species, or that interactions are not size structured. Indeed, a parasite outbreak in Michigan streams demonstrated a large, relatively unappreciated role for strong species interactions, including competition (Kohler and Wiley 1997); the ecosystem engineering of beavers in rivers is characteristic of a keystone species (**Naiman et al. 1986** [chapter 9]); and introduced trout have completely changed trophic interactions in New Zealand streams (see chapter 5). However, the distinct conditions imposed by a running-water environment result in communities and interactions that are unique to streams (see chapter 3). Consideration of the key thematic developments that shaped our understanding of species interactions and communities in streams from the 1960s to the 1980s yields three main insights about the underlying drivers of stream community structure.

First, stream ecologists have long appreciated that flow-related dispersal is a fundamental aspect of movement and interaction of biota in streams, initially through the intensive study of the movement of stream macroinvertebrates in the water column (i.e., drift), as highlighted in chapter 1 as well as in **Waters (1962)**. Key developments in dispersal studies included documenting the causes and consequences of drift (Townsend and Hildrew 1976; Kohler 1983), the magnitudes and distances traveled (Elliott 1965, 1971), the drivers that shape its periodicity and composition (Holt and Waters 1967; Allan 1978; Flecker 1992), and its role in influencing the food supply for stream predators (Elliott 1970; Fausch 1984).

A second thematic advance in stream community ecology was a recognition that the advective stream environment may obscure the influences of species interactions on populations and communities, including predator-prey interactions (**Peckarsky 1980**), herbivory (**Lamberti and Resh 1983**), and competition (**McAuliffe 1984**). For example, cage and fence experiments, including some ambitious large-scale field manipulations (e.g., Allan 1982), indicate that macroinvertebrate movement and fast turnover can make the local effects of high predation rates on invertebrates by fish difficult to detect (leading to "Allen's paradox"; Allen 1951; see chapter 5). Cooper et al. (1990) showed that outcomes of cage experiments assessing predation effects were linked to movement: small-mesh cages that impeded invertebrate movement were much more likely to result in prey declines than were controls with larger mesh. A more nuanced understanding of how the advective stream environment affects species interactions subsequently developed, which recognized important roles for scale (Englund 1997; Anderson et al. 2005), nonconsumptive and behavioral effects (Peckarsky et al. 2002), and the effect of variations among taxa and life stages (Lancaster et al. 2010; Lancaster and Downes 2018).

The third and final theme reflects the role of disturbances, especially floods and droughts, and their effects on stream environments, the importance of interactions, and the assembly of stream communities. The physical constraints associated with flow-related disturbance regimes produce a set of multi-scale environmental filters that determines "who" can live and compete in any given location (Resh et al. 1994; **Townsend and Hildrew 1994** [chapter 3]; Poff 1997). Environmental harshness associated with disturbances clearly has a large influence on the composition of stream communities (Fisher et al. 1982; Death and Winterbourn 1995; **Stanley et al. 1997** [chapter 3]). This fact caused stream ecologists to consider how disturbances may moderate the outcomes of species interactions (Peckarsky 1983; Townsend et al. 1997) and the processes involved in recolonization and succession after disturbance (Fisher et al. 1982), as well as the patchy mosaic of habitats created and the roles of deterministic relative to stochastic processes in driving community assembly (Pringle et al. 1988; **Townsend 1989**).

## HIGHLIGHTED PAPERS

Waters, T. F. 1962. "Diurnal Periodicity in the Drift of Stream Invertebrates." *Ecology* 43:316–20.

Thomas F. Waters's 1962 paper was foundational in establishing the magnitude and regularity of downstream movement, or "drift," by stream invertebrates. We now know that the occurrence of drift (figure 4.1), in

Fig. 4.1 A drifting New Zealand mayfly, *Deleatidium* sp., showing the typical behavior in which legs and caudal filaments (i.e., "tails") are spread while moving in the water column. Photo courtesy of A. R. McIntosh.

which a stream organism leaves the streambed and travels downstream in the water column, is one of the more important processes in stream community ecology, with far-reaching consequences for population dynamics and community assembly (Brittain and Eikeland 1988). Drift also forms the food supply for many stream predators, especially "drift-feeding" fishes, and so represents an important feature of many stream food webs (Elliott 1973).

Although the occurrence of drift was described anecdotally around the world as early as the 1920s (e.g., Needham 1928), quantitative evaluation did not begin until the 1960s, after European scientists started measuring its characteristics (Müller 1954, 1974). Waters's study of the diurnal periodicity of drift in Valley Creek, Minnesota, USA, was pivotal because it measured drift hourly over twenty-four hours, highlighting not just the nocturnality of the movement but also its high magnitude.

Waters's development and refinement of the sampling approach played an important part in facilitating standardized quantitative measurements of aquatic invertebrate movement, although modern drift density units account for flow differences by including water velocity measurements (Allan and Feifarek 1989). Waters produced an hourly time series of standardized drift-net samples that revealed remarkable patterns of nocturnal drifting in mobile macroinvertebrates such as the amphipod *Gammarus*, the mayfly *Baetis*, and the cased caddisfly *Glossosoma*. Besides being higher at night than during the day, drift rates often had large post-dusk peaks, were suppressed by moonlight, and almost completely ceased after dawn. Children playing upstream in the study stream one afternoon also provided an unintended, but valuable, ad hoc experiment: their disturbance increased the daytime drift of mobile taxa when there had otherwise been none, but did not affect the larvae of *Glossosoma*, which were presumably able to regain their positions quickly due to their heavy cases. The nocturnality in drift was linked to predation risk from predatory fish, leading to a large body of subsequent work describing how drift-feeding fish influence patterns of drift (e.g., Flecker 1992).

The most important revelation from Waters's paper had to do with the magnitude of drift. Previous studies had not measured drift at night, or had lumped night and day, but his study showed that the number of individuals moving per unit time was enormous (e.g., up to thousands of organisms per hour), particularly just after dusk. In fact, Waters showed that daily drift rates over an area of streambed could be many times the standing stock of that area. Given the often high magnitude of invertebrate drift, a long-debated question became why upstream populations were not depleted, leading to what became known as the "drift paradox" (Müller 1982). A combination of density dependence, random dispersal, and the high secondary production of benthic habitats likely explains the persistence of populations (Anholt 1995; Naman et al. 2016), but many questions still remain about the scales at which those processes operate (Anderson et al. 2005).

Williams, D. D., and H. B. N. Hynes. 1976. "The Recolonization Mechanisms of Stream Benthos." *Oikos* 27:265–72.

As the work of Waters highlights, the downstream movement of macroinvertebrates by drift is a foundational concept in stream ecology, perhaps as a result of its uniqueness to these unidirectionally flowing systems. Because floods are also common in streams, drift was thought to be a critical pathway of recolonization following these disturbances (for both invertebrates and plants; Nilsson et al. 2010). Thus, Williams and Hynes built on work from the 1950s to 1970s by Müller, Elliott, Waters, and others to quantify the contribution of downstream drift to the recolonization of the stream benthos relative to three other sources: upstream migration, upward migration from within the substrate, and aerial oviposition. While it had already been observed that the benthos recolonized rapidly following disturbances (Waters 1964), this study established the relative contributions of these different pathways.

To differentiate the four recolonization pathways, the authors devised a novel system using four different wooden traps, each open to one recolonization source, placed in a small stream draining an agricultural landscape in Ontario, Canada, for twenty-eight days. Drift was the dominant source of recolonization, contributing approximately 41% overall, relative to 28% from aerial sources, and 18% and 19% from upstream migration and upward movement, respectively. However, rates of recolonization by the four pathways differed substantially among taxa. For instance, bivalves were almost entirely restricted to upstream movement, whereas mayflies and caddisflies tended to drift more, and beetles were equally likely to recolonize via drift, upstream movement, or upward movement from the hyporheic zone (i.e., the interstitial spaces below the streambed; see chapter 2).

By breaking recolonization into these four distinct pathways, Williams and Hynes's research became foundational in promoting more spatially oriented stream research, despite the findings having come from one small agricultural stream system. For example, their paper highlighted the potential role of areas like the hyporheic zone, which have since figured strongly in establishing a more complete understanding of stream environments. Thus, their work laid the theoretical groundwork for many further observational and experimental studies on stream recolonization (e.g., Fournier et al. 2023). Having a greater appreciation of these recolonization pathways also paved the way for future research exploring the role of dispersal and interchange among different freshwater locations, from patches to riverscapes, along multiple dimensions. For instance, **Ward's (1989** [chapter 3]) four-dimensional perspective on running waters considered the important interchanges along various dimensions: longitudinal, lateral, vertical, and temporal. The spatiotemporally dynamic nature of stream ecosystems is characterized by these various interconnections.

We now recognize that connectedness across multiple pathways is an integral part of the character and resilience of streams and rivers. For instance, Hauer et al. (2016) emphasized how the landscape connections of gravel-bed rivers in all four dimensions, including the considerable interchanges of organisms between groundwater and surface water, underpin the provision of resources to a wide range of aquatic and terrestrial biota. Situations in which restoration of abiotic stream conditions fails to result in sought-after biotic shifts may be linked to a lack of recruits via one or more colonization pathways (Palmer et al. 2014). Finally, the recent focus on deconstructing the importance of dispersal via different pathways relative to environmental selection has emphasized the role of those connections in structuring metacommunities (Tonkin et al. 2018).

Peckarsky, B. L. 1980. "Predator-Prey Interactions between Stoneflies and Mayflies: Behavioral Observations." *Ecology* 61:932–43.

In the introduction to her classic 1980 paper, Bobbi Peckarsky observed that there was abundant evidence for predator influences on the communities of aquatic ecosystems other than streams and rivers, including ponds, lakes, and marine rocky intertidal ecosystems, but that such evidence was rare for streams. This paper marked the beginning of a forty-year research program led by Peckarsky at the Rocky Mountain Biological Laboratory in Colorado, USA (see chapter 10), which revealed how predators influence stream communities and how these effects differ from those that are characteristic of other ecosystems. Streams are fundamentally open systems in which complex life cycles and the advective environment result in organisms regularly moving in and out of local patches. Peckarsky's early experiments advanced our understanding of how this movement shapes predator-prey interactions.

Peckarsky aimed to determine how mayfly nymphs detected and avoided predatory stoneflies. Movement was a common outcome of stonefly encounters for some mayfly taxa, especially the widespread *Baetis*. The dramatic nocturnal drift of many stream macroinvertebrates, like that observed in **Waters's (1962)** study, had been assumed to be a consequence of nighttime positioning on upper surfaces of rocks, where accidental dislodgement was more likely. However, this study revealed that mayflies had much more control of their movements than had been previously appreciated, and that drifting could be an active predator avoidance response. In comparison to *Baetis* mayflies, the better-armored *Ephemerella* mayflies took up a scorpion-like posture in response to stonefly threats, a strategy that very rarely led to drift. Thus, this experiment showed that predators considerably altered the positioning, movement, and distribution patterns of their prey, and that prey exercised a considerable degree of control over their movements. The results also highlighted the importance of chemical signals transported in the water column for predator avoidance: none of the prey responded to visual stimuli only. Flowing water moves and conserves chemical cues (Dodson et al. 1994), providing a directional concentration gradient that triggers responses involving changes in movement and patterns of development (Peckarsky et al. 2002).

Such knowledge of predator effects on prey movements has been instrumental in understanding the influence of predators on stream communities. The observed small-scale responses to predatory invertebrates like stoneflies show that mobile prey like *Baetis* are more likely to leave patches with invertebrate predators than control patches lacking these predators, accentuating the apparent effect of the predators on prey abundance. By comparison, the opposite occurs for predators like drift-feeding fish, which suppress movement of mobile prey like *Baetis*. Thus, movement, and treatments that affect it, like cages, greatly influence the outcomes of predator manipulations (Cooper et al. 1990), including their dependence on the scale of movement in relation to the spatial scale of experimental manipulation (Englund 1997). Collectively, these movements and the life-history responses of prey help explain why stream predators can consume large amounts of benthic prey while still not depleting prey populations.

Lamberti, G. A., and V. H. Resh. 1983. "Stream Periphyton and Insect Herbivores: An Experimental Study of Grazing by a Caddisfly Population." *Ecology* 64:1124–35.

The first sentence of Lamberti and Resh's 1983 paper reads, "Most advances in our knowledge of the interactions between herbivores and algae have resulted from studies in the marine environment" (p. 1124). Their study was one of the first to examine those interactions in streams experimentally. Lamberti and Resh's experiment revealed how the characteristics of a running-water environment could influence herbivory and fuel algal-driven productivity in streams.

Their study focused on *Helicopsyche* caddisfly larvae, the dominant grazer in Big Sulphur Creek in Northern California, USA, which is a distinctive stream macroinvertebrate with a heavy, spiral-shaped larval case. The experiment involved the ingenious idea of using artificial substrates (clay tiles) raised off the streambed in pools to create a grazer "exclusion" treatment for comparison with "grazed" plots on the streambed (figure 4.2). The heavy-cased *Helicopsyche* could not swim to the raised platforms, which showed dramatic increases in the algal standing crop. Lamberti and Resh's innovative method for creating a grazer exclusion treatment was important because it facilitated an experimental comparison that had typically relied on cages in other ecosystems, which would have been problematic in streams because of the disruption cages caused to water flow. Many other studies aiming to test various controls over stream community structure have subsequently devised a variety of clever approaches to solve this problem. These approaches have ranged from the extensive use of artificial stream channels (e.g., Crowl and Covich 1990; Rosemond et al. 1993, 2001) to the

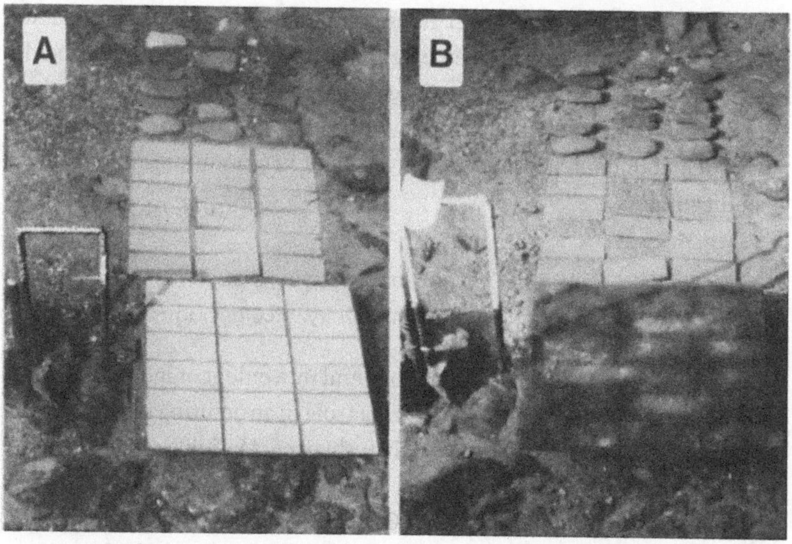


Fig. 4.2 Exclusion (elevated tiles) and grazed (tiles directly on substrate) plots in Big Sulphur Creek (A) at the beginning of the grazer-exclusion experiment (September 5, 1980) and (B) after five weeks (October 11, 1980). Note the abundant filamentous green algae on exclusion plot tiles. The rocks above the control tiles were used to calibrate the sampling effectiveness of the artificial substrates. Direction of water flow is from right to left. Tiles and rocks missing in (B) are those sampled just before the photograph was taken. From Lamberti, G. A., and V. H. Resh. 1983. "Stream Periphyton and Insect Herbivores: An Experimental Study of Grazing by a Caddisfly Population." *Ecology* 64:1124–35.

deployment of underwater electric fences (e.g., Pringle and Blake 1994). Lamberti and Resh's use of raised platforms in pools to exclude a relatively immobile grazer was unusual, given that many stream inhabitants are proficient swimmers, but it again emphasized the importance of differences in stream macroinvertebrate mobility.

In addition to revealing dramatic increases in stream periphyton abundance when grazers were excluded, Lamberti and Resh's experiment emphasized the potential productivity of grazed stream periphyton. The authors measured algal productivity using oxygen consumption in light and dark chambers, detailed the composition of algal assemblages, and even measured bacterial abundance. Their findings showed that a low-biomass, grazed "lawn" of diatoms could still result in primary production sufficient to support a high biomass of herbivores, probably due to high turnover rates combined with reduced light and nutrient limitation. Thus, they essentially revealed how grazers could accelerate stream productivity. These findings still provide one of the most important explanations for how stream autotrophic production fuels such abundant stream macroinvertebrate communities in shallow, sunlit streams.

McAuliffe, J. R. 1984. "Competition for Space, Disturbance, and the Structure of a Benthic Stream Community." *Ecology* 65:894–908.

The early work on species interactions and movement of individuals set the scene for understanding their wider role in structuring stream communities, especially in the face of disturbances. McAuliffe's 1984 study examined the role of competition and disturbance in structuring the benthic insect community in Owl Creek, a small lake-fed stream near Missoula, Montana, USA. This work came during a period of intense debate in the ecological literature over whether biotic interactions held communities at some form of equilibrium or if disturbance kept communities in a perpetual state of recovery from the last disturbance (Connell 1978; Huston 1979). Streams were typically considered to be at the nonequilibrium end of this spectrum, given their physically dynamic nature (Fisher et al. 1982; Grimm and Fisher 1989). Stream ecologists now appreciate the myriad processes that interact to determine the structure of ecological communities, and they understand that these processes operate at multiple scales (Vellend 2010; Thompson et al. 2020) and with a high level of context dependency, making the identification of general rules challenging (Lawton 1999). However, McAuliffe's study kept the debate over the strength of equilibrium versus nonequilibrium processes alive for streams by demonstrating how influential interactions could be in determining the structure of a benthic community.

Over a two-year period, McAuliffe measured the evenness of communities in relation to disturbance, observed microdistributions via photographs or tracings of sessile invertebrates on stones, and undertook a suite of experimental manipulations to tease apart interactions between species. The shelter-building caddisfly *Leucotrichia pictipes* was particularly influential, as it negatively affected several other sessile species, including the chironomids *Eukiefferiella* sp. and *Rheotanytarsus* sp. and the moth *Parargyractis confusalis*. *Leucotrichia* aggressively defends space around its retreat and, with enough time, competitively monopolizes the surfaces of stones, reducing the abundance of, or even completely excluding, other species. However, *Leucotrichia* does not respond well to disturbance, whether via flooding or drying.

The dynamics of community assembly were clearly apparent in the experimental results, which showed that the potential for stochastic colonization determined the community trajectory via priority effects, and that disturbance reduced the competitive dominance of *Leucotrichia*. *Parargyractis* arriving first made the habitat unsuitable for *Leucotrichia*, which was otherwise capable of preventing all others from establishing. Thus, timing was important in this community, in which most colonizers arrive within a few weeks in late summer. However, *Leucotrichia* larvae can use retreats built by previous generations, which can persist from year to year. Having this head start over *Parargyractis*, which must build retreats each year, gives *Leucotrichia* a competitive advantage. Removing *Leucotrichia*, therefore, not only allows other sessile species to increase in number, but also makes way for disturbance to influence community trajectories. The time required to build new retreats leaves *Leucotrichia* at a disadvantage when stones overturn during floods, destroying retreats, and thus gives *Parargyractis* the advantage on less stable substrates. Moreover, shorter-lived *Eukiefferiella* benefited from prior drying of large stones, as their life cycle, with several generations per year, enabled them to occupy newly available stone surfaces following rewetting. These patterns were revealed only by a remarkable level of manipulation and observation over sustained periods, producing insights that continue to influence stream ecology.

The study highlights the importance of considering both the spatial and temporal dynamics of ecological processes in streams to understand community dynamics. As Hutchinson's (1967) "paradox of the plankton" demonstrated for lakes, several sessile stream organisms can coexist in a patchy and dynamic mosaic of conditions facilitated by different rates of disturbance. In the absence of disturbance, *Leucotrichia* dominates communities. Many stream ecologists regard disturbance as a dominant organizing force of communities. However, in ecosystems that are generally considered disturbance driven, the clear outcome of competition in McAuliffe's paper serves as a reminder that biotic interactions can indeed be extremely important in structuring stream communities.

Townsend, C. R. 1989. "The Patch Dynamics Concept of Stream Community Ecology." *Journal of the North American Benthological Society* 8:36–50.

If ever a paper in stream ecology was ahead of its time, it was Colin Townsend's foundational work on patch dynamics in streams. Townsend's 1988 plenary lecture to the North American Benthological Society annual meeting, which gave rise to this paper, covered topics that remain in vogue in stream community ecology as a result of the surge in research on metacommunity structure and dynamics. Townsend set out to evaluate the current state of understanding and knowledge gaps in stream community ecology, and in doing so laid out a compelling framework that emphasized the roles of patch heterogeneity, stochastic events, and animal movement. Townsend argued that the concept of patch dynamics may provide a more general basis for understanding stream community structure and dynamics than other contemporary concepts, such as the RCC (e.g., **Vannote et al. 1980** [chapter 3]). He suggested that stream communities remain best characterized as highly dynamic, reflecting considerable temporal variability in conditions, where history, chance, disturbance, and colonization play key roles in structuring assemblages, and where species that may be considered weedy are widespread.

Townsend's perspective has had a significant impact on the study and management of stream ecosystems. His conceptual framework helped to synthesize various views of how stream communities are structured and how they respond to disturbances, particularly given the patchy distribution and temporal dynamism of their physical and biological features. He emphasized the importance of disturbance as a principal driver of community structure in streams, reflecting the significant attention to this concept in the literature at the time. Townsend's conceptual packaging of the interactions between stochastic and deterministic processes in streams, as well as the importance of refugia, dispersal, and colonization, into a coherent framework represents a master class in synthesizing across theory and a broad swath of empirical research.

Townsend also aimed to clarify where streams fit in with respect to the major drivers of community structure, seeking an understanding of the relative influences of deterministic (e.g., species interactions) versus stochastic (e.g., flow-driven disturbances) processes, and of the importance of considering changes in biotic and abiotic conditions over time (i.e., the temporal dimension). Townsend predicted that temporal variation in streams could allow greater diversity via niche segregation of similar species among seasons, and this prediction found subsequent support (Benke and Jacobi 1994; Tonkin et al. 2017). The need to consider time more robustly is a repeating theme in community ecology due to its context dependency, with some going so far as to say that "community ecology is a mess, with so much contingency that useful generalizations are hard to find." (Lawton 1999, p. 178). This contingency is particularly important in ecosystems with temporal forcings as strong as those in streams, yet we remain starved of time-series data in stream ecology to adequately examine these ideas, likely due to the costs associated with the necessary data collection.

Townsend also argues that the strongest role for species interactions in streams often emerges among species that are largely sedentary and competing for space (e.g., **McAuliffe 1984**). Yet he points out that most stream invertebrates are highly mobile, with diverse tolerances and requirements, making frameworks like patch dynamics helpful in understanding community-level patterns in disturbance-prone streams (Pringle et al. 1988). Indeed, patch dynamics and other spatially focused frameworks are particularly relevant, as Townsend points out, when considering the rates of dispersal of stream insects, particularly via downstream drift. The continuous loss of animals from the substrate to the drift, and the associated continuous colonization by new recruits, plays a major role in structuring stream communities. Townsend notes that such high rates of colonization and drift can potentially induce "local species interactions to be swamped and . . . community composition, at least on the small scale, to be largely determined by disturbance and colonization dynamics within a patchy environment."

## LOOKING FORWARD

One of the common threads linking the studies highlighted in this chapter is their relatively narrow spatial scale: the empirical studies we have highlighted all involve work in a single study reach or site. In contrast, stream ecologists focusing on fish were early to recognize the importance of dispersal across the whole interconnected dendritic network of branches that characterizes streams, and encouraged stream ecologists to take a larger, "riverscape"-scale view of dispersal in streams (**Schlosser 1991** [chapter 3]; Fausch et al. 2002). One of the most important recent developments from the 1990s onward has been the increasing expansion of the spatial and temporal scales at which stream ecologists work. This expansion has been particularly important given widespread human alterations to streamflow and conditions (Stewart et al. 2005; Blöschl et al. 2017) and propagation of those human influences via stream networks (Poff et al. 2007). Thus, a view of stream communities as occupying many interconnected locations, and as being influenced by large-scale dispersal between them, underpins the current understanding of the role of species and community interactions across stream networks and how we conceptualize the assembly of what we now call stream metacommunities (Campbell Grant et al. 2007; Tonkin et al. 2018).

Given the expanded spatial scales and huge datasets to which stream ecologists now have access, one might be tempted to dismiss the work we highlight here as small-scale and of only historical relevance. However, modern stream ecologists might do well to take heed of the lessons of the past. **McAuliffe (1984)** ends his paper by emphasizing that if we want to detect the role of biological interactions in streams, we need to adjust the scale at which we study them, arguing that we should be observing these processes on the scale at which organisms influence each other. For example, despite known strong interactions between *Leucotrichia* and *Parargyractis*, there was no indication of the strength of those interactions in the observational data. This is an important finding that needs to be considered in any modern study of large-scale species distributions: most mass-sampling techniques are insufficient to pick up such interspecific interactions. Thus, there are lessons for the present-day quantitative ecologists employing methods, whether appropriate or not, to detect interactions from large-scale distribution data. Stream ecologists remain ill equipped to explain the variability in stream community structure, despite massive advances in technology and statistical techniques. A challenge going forward is to develop strong-inference approaches relevant to all the processes determining stream community structure, and at the spatial scales at which stream ecologists desire to work. In addition, we must not forget the importance of basic natural history information, which is fundamental to many modeling approaches that seek to forecast the responses of stream communities to future threats (Tonkin et al. 2019). The field skills deployed by the studies highlighted in this chapter remain as relevant today as they were when those studies were undertaken.

Drift was a foundational theme of stream ecology that was very much about natural history discovery. **Williams and Hynes (1976)** uncovered much about the various sources of recolonization of the stream benthos. However, we still know relatively little about the adult stages of stream insects that disperse in the terrestrial environment. Despite identification of the importance of the adult, oviposition, egg, and early life stages in the 1950s, we have still largely failed to confront the full population cycle, and studying the population ecology of stream organisms has gone out of vogue. This decline in research effort has left large and critical gaps that cannot be filled by more correlative analyses of community composition data. Resolution of this issue involves consideration of underappreciated influences on stream macroinvertebrates throughout their life cycles, including the dispersal of adult terrestrial stages (Lancaster and Downes 2018), the role of density dependence (Anholt 1995; Hildrew et al. 2004; McIntosh et al. 2022), and the potential for recruitment limitation (Lancaster et al. 2010; Encalada and Peckarsky 2012).

Finally, one of the biggest remaining challenges for stream ecologists is to embrace fully the plurality of processes that structure stream communities. Much stream ecology work from the 1980s and 1990s grappled with the relative influences of deterministic and stochastic forces in shaping communities. While modern metacommunity ecology incorporates a large variety of processes that are important in assembling stream communities and better defines the scales at which they operate (Vellend 2010; Thompson et al. 2020), we have still tended to pigeonhole streams as being driven by particular types of drivers. If the thirty-five years since the publication of **Townsend's 1989** patch dynamics paper have taught us anything, it is that we need to consider the complex interplay between processes at different spatial and temporal scales, including species interactions, to advance understanding of the assembly of stream communities.

# 5

# FOOD WEBS AND TROPHIC INTERACTIONS

Alexander D. Huryn and Alan G. Hildrew

Ecological communities are groups of potentially interacting species. Many of these interactions involve "feeding"—grazing, predation, parasitism, and disease. If we catalog and record them, we can depict a "food web" (**Hildrew et al. 1985; Winemiller 1990**; Woodward, Speirs et al. 2005; Woodward, Thompson et al. 2005). Species interactions are important in terms of both population dynamics and flows of energy and materials through the community, and food webs thus link both great traditions of ecology: population/community studies and ecosystem studies (**Benke and Wallace 1980; Hildrew et al. 1985; Power et al. 1985; Huryn 1998**). Early attempts at mapping trophic interactions in streams revealed complex food webs (Jones 1950; **Cummins et al. 1966; Minshall 1967**). In general, these were qualitative webs, showing connectedness ("connectance") rather than the strengths of interactions. At its simplest, connectance is the fraction of all possible links observed at least once during the study. Connectance is fiendishly difficult to measure, however, and depends on factors such as taxonomic coverage and resolution (i.e., how complete our sample of the community is and how finely the taxa are identified; **Winemiller 1990**; Thompson et al. 2001). Construction of a well-resolved food web is therefore laborious, but easier if the web is simplified by grouping species with similar feeding traits into "functional feeding groups" (**Cummins et al. 1966**; Cummins 1973). Such groups contain organisms that obtain their food in a particular way—for instance, animals that can chew large particles ("shredders"). These guilds do not imply a particular diet (e.g., Tierno de Figueroa 2019), though the assumption that they do has provided much of our current "textbook" view of stream ecology, and particularly of invertebrate feeding interactions.

A primary goal of studying food webs is to predict how their structure affects trophic interactions and ultimately ecosystem processes. Consequently, many recent food-web analyses extend beyond simple connectance by measuring energy and material flows between taxa—the "nodes" of the web (Benke 2018). An early example of how trophic interactions relate to ecosystem processes was described by Allen (1951), who found that his estimates of prey production in a New Zealand stream were apparently insufficient to support observed demand by brown trout. While this "Allen paradox" has now been resolved (Waters 1988; **Huryn 1998**) such estimates of predator impact, based on comparisons of predator and prey production, do often indicate severe top-down effects through the web. It is thus not surprising that some studies have detected "trophic cascades" in stream ecosystems, in which direct and indirect effects of predation on herbivores affect algal accrual (**Power et al. 1985; Huryn 1998**).

## HIGHLIGHTED PAPERS

Minshall, G. W. 1967. "Role of Allochthonous Detritus in the Trophic Structure of a Woodland Springbrook Community." *Ecology* 48:139–49.

Cummins, K. W., W. P. Coffman, and P. A. Roff. 1966. "Trophic Relationships in a Small Woodland Stream." *Internationale Vereinigung für theoretische und angewandte Limnologie: Verhandlungen* 16:627–38.

Minshall (1967) and Cummins and colleagues (1966) influenced the development of modern stream ecology substantially, particularly through their emphasis on the importance of imported ("allochthonous") detritus and generalist diets, which remain foci of research today. Readers are encouraged to imagine what it would be like to explore stream ecosystems as a graduate student at the University of Louisville (G. W. Minshall) or an assistant professor at the University of Pittsburgh (K. R. Cummins) during the 1960s, when many of their functional attributes, taken for granted today, were largely unexplored. Although these two studies were conceived independently, they are similar in scope and are usefully treated together. Deceptively simple, they provide insight into a moment when modern conceptualizations of the major trophic pathways in stream food webs were emerging. Reading them, one senses the origins of modern concepts such as (1) the quantitative dominance of detritus within the energy base of forested stream ecosystems, (2) the explicit recognition of "green versus brown" energy channels through food webs (Zou et al. 2016), and (3) the importance of omnivory (feeding at more than one trophic level), as well as the need for well-resolved food webs to assess such concepts. These studies also contributed widely used methods, including the quantification of detritus, the simplification of food webs by combining species by function, and the use of filtration and membrane filters for dietary analysis.

Minshall studied Morgan's Creek, a small, forested headwater stream in Kentucky, USA. His goal was to describe the trophic structure of a stream community by year-round sampling of the biomass of plant materials available to consumers as well as invertebrate abundances and diets. To his knowledge, the only other study that focused on the role of detritus in streams was by **Nelson and Scott (1962** [chapter 1]), who studied a bedrock-outcrop community in the Oconee River of Georgia. Minshall showed that detritus in the form of autumn-shed tree leaves was the primary food ingested by invertebrates. Analysis of gut contents and of principal pathways of energy flow in the stream using a connectance food web indicated that autumn-shed leaf detritus provided the main source of food for primary consumers (50%–100% of food ingested), and Minshall inferred that this detritus was a major (indirect) energy source for higher trophic levels.

The 1966 study by K. W. Cummins and coauthors focused on Linesville Creek, a small, forested headwater stream in northwestern Pennsylvania, USA. Their goal, like Minshall's, was to assess the "trophic relations" of the entire community of riffle-inhabiting invertebrates. Their study was an early attempt at characterizing trophic structure by producing a food web for a stream community. The gut contents of invertebrates were drawn onto membrane filters and quantified under a microscope, a technique still widely used today (Mecom and Cummins 1964; Rosi-Marshall et al. 2016). The resulting food web was thought to be too complex for analysis and was simplified using dietary analysis to assign different nodes to specific trophic levels: detritus, detritus consumers, grazers of living primary producers, living primary producers, and predators. This procedure foreshadowed the development of the functional feeding group concept by Cummins (1973). The conceptual flow chart used to visualize the ecosystem (figure 5.1) illustrates the early quantitative recognition of the tangle of green and brown channels in freshwater food webs. The authors' take-home message, similar to Minshall's, was that much of the benthic fauna consists of dietary

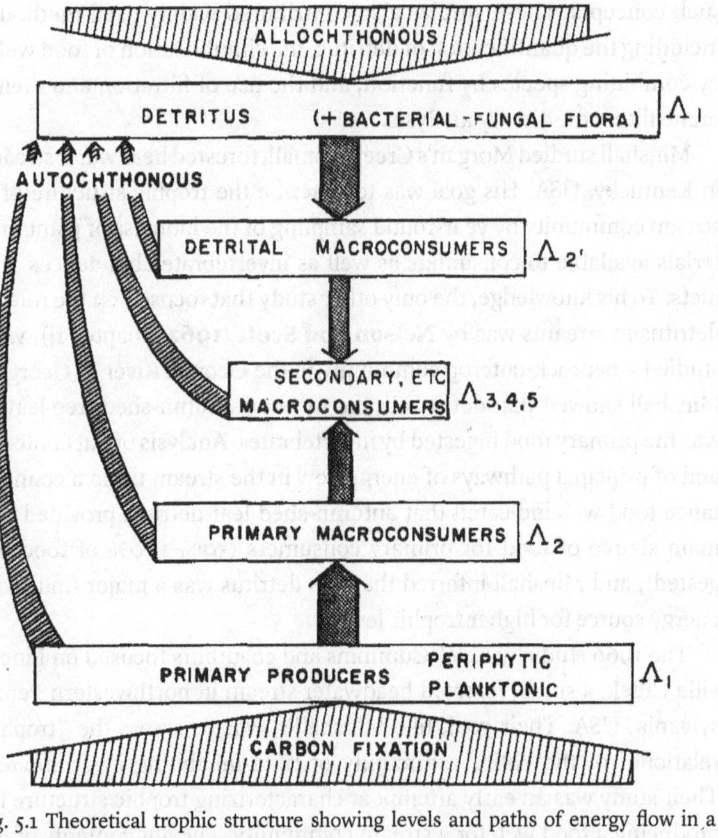


Fig. 5.1 Theoretical trophic structure showing levels and paths of energy flow in a woodland stream ecosystem. From Cummins, K. W., W. P. Coffman, and P. A. Roff. 1966. "Trophic Relationships in a Small Woodland Stream." *Internationale Vereinigung für theoretische und angewandte Limnologie: Verhandlungen* 16:627–38. Reprinted by permission of Taylor & Francis Ltd.

generalists, and that detritus has a prominent quantitative role in their diet. The proposition that terrestrially sourced detritus dominates the energy base for forested headwater stream communities, as championed by Hynes (1963), was clearly supported.

Benke, A. C., and J. B. Wallace. 1980. "Trophic Basis of Production among Net-Spinning Caddisflies in a Southern Appalachian Stream." *Ecology* 61:108–18.

A novel and highly influential study by Benke and Wallace explored the specific pathways and quantities of organic matter flow through a group of filter-feeding caddisflies. Although the omnivorous feeding behavior of this group had been well documented (e.g., Hildrew and Edington 1979), its role in specific energy pathways was poorly understood because feeding studies had not been integrated with material flow analyses. The results of Benke and Wallace's 1980 study showed for the first time that animal prey supported the greatest proportion of the production by this functional feeding group of caddisflies (80%), even though detritus was the dominant food ingested. As perhaps the first study to quantify the role of caddisflies in the "spiraling" of seston (suspended particles) in streams, it showed that the mass of detritus released as fecal pellets by filter-feeding caddisflies, as a group, was about the same as the mass they consumed. However, these caddisflies also effectively decreased the food quality of the seston. Although this study was a significant advance in understanding the community energetics of filter-feeding insects at the time, its results were probably less influential than its introduction of the "trophic basis of production" approach (Benke and Huryn 2017).

The story behind the conceptualization of the trophic basis of production approach provides an example of how networking among researchers with different perspectives and personalities can expand the intellectual scope of a particular study. The core data underpinning the 1980 study described annual patterns of abundance and diets of six species of filter-feeding caddisflies in a North Georgia mountain stream. These data were collected by J. B. Wallace (University of Georgia, Athens), who at the time was deeply interested in filter feeding and seston spiraling (Wallace et al. 1977). After attending a 1979 seminar where Wallace presented his data, A. C. Benke (then at the Georgia

Institute of Technology) approached him to discuss the fact that his data contained the elements required to estimate the production of the caddisflies. At that time, Wallace had no experience with production ecology and thought Benke's idea would provide an opportunity for further exploration. While at dinner that evening, Benke sketched an outline of an analytical approach that could be used to integrate Wallace's quantification of the caddisflies' diets with estimates of their production and literature-derived ecological efficiencies to produce a flow food web allowing the quantification and visualization of the pathways and quantities of material flows through the filter-feeding caddisfly community. This collaboration led to the elegantly simple "trophic basis of production" approach, which relies on the integration of food-specific assimilation and production efficiencies and quantitative descriptions of diet and secondary production. At the time of its publication, Benke and Wallace's study represented the cutting edge of stream bioenergetic research, and the resulting methodological approaches have been used up to the present day (Benke and Huryn 2017).

Hildrew, A. G., C. R. Townsend, and A. Hasham. 1985. "The Predatory Chironomidae of an Iron-Rich Stream: Feeding Ecology and Food Web Structure." *Ecological Entomology* 10:403–13.

The 1970s and 1980s saw a renewed interest in food webs from a theoretical point of view, and particularly in features affecting community dynamics—relating to the long-debated question of whether high species diversity promotes or reduces community stability (Elton 1927; May 1973). To address this question, theoreticians compiled data on real food webs from a variety of ecosystems, though most of these data had been collected for other purposes and were inadequate in this new context. Hildrew and Townsend (1976) had begun their series of studies of the benthos of the acidic, iron-rich, and fishless Broadstone Stream, which was attractive due to its apparent simplicity and dominance by a few abundant, large-bodied invertebrate predators and a limited assemblage of midges and stoneflies. They saw it as a natural community, yet one that was sufficiently tractable to unravel species interactions within a multispecies context. Their 1985 study was aimed originally at a straightforward description of the diets of the three species of smaller-bodied "predatory" midge larvae. The authors resolved these diets as far as they could and, combining their findings with data on the larger predatory species, constructed a simple food web—which nevertheless turned out to be among the most detailed webs then available, certainly for a whole freshwater community. It was at first a connectance web; that is, it simply recorded the occurrence of links and made no attempt to assess their strength.

Some features of this first Broadstone Stream web immediately seemed unusual. In comparison to other food webs then available, its connectance was high—a feature at that time theoretically associated with a "fragile" and dynamically unstable system (Briand 1983) and apparently associated with a constant environment (which is unlikely for any real stream). There was also a great deal of omnivory—"omnivory" being defined as feeding at more than one trophic level, not as simple dietary generalism. Omnivory, too, was theoretically expected to reduce web stability.

Overall, this study revealed a complex web of life-cycle (ontogenetic) omnivory, in which diet changes with size and development, and a pattern of predator-prey reversals based on relative body size, in which two species are vulnerable to each other at different life-cycle stages. It also pointed to predatory links with other, even smaller metazoan animals (the "meiofauna"; Schmid-Araya and Schmid 2000), indicating that even this web was an oversimplification. It confirmed the need for better food-web data, preceded influential studies from other systems confirming the divergence of real food webs from contemporary theoretical expectations (Polis 1991), and led to a "flowering" of food-web studies (Closs and Lake 1994; Tavares-Cromer and Williams 1996) as well as to further research on Broadstone Stream demonstrating the unsuspected complexity of this, and other, real food webs (Schmid-Araya et al. 2002).

Many of the early theoretical expectations about food webs have since been greatly modified, and food-web ecology has moved on to consider linkage strength (Woodward, Speirs et al. 2005; Woodward, Thompson et al. 2005; Benke 2018), rather than the simple occurrence of links, whereas the key role of body size in structuring stream food webs is now clear (Woodward and Warren 2007).

Power, M. E., W. J. Matthews, and A. J. Stewart. 1985. "Grazing Minnows, Piscivorous Bass, and Stream Algae: Dynamics of a Strong Interaction." *Ecology* 66:1448–56.

Power, Matthews, and Stewart's 1985 paper was one of a few that bridged a gap between research focused on single species or pairs of species and the study of more complete food webs by adopting an approach based on an intimate knowledge of natural history strengthened by field experiments. It identified strongly interacting subsets of food webs (small groups of species, now sometimes called "modules" or "motifs"; Stouffer et al. 2007) and showed how the effects of such interactions ramified through a web (via indirect interactions) to produce a trophic cascade. It also showed nonconsumptive effects of predators via the responses of prey to the threat of predation. It even indicated the effects of flow disturbances and the cessation of processes based on biotic interactions when flow was high. Overall, this paper was an elegant and fascinating demonstration of the strength of field experiments, and Mary Power was one of the founders of the productive use of field experiments in stream ecology.

The setting for Power and her colleagues' study was a stream in Oklahoma, USA, during prolonged periods of low flow in summer and autumn, where there were piscivorous bass, a grazing minnow, and an assemblage of attached algae. The team noticed that, in replicate pools spread along a short reach of the stream, the densities of minnows and bass were almost always inversely correlated, a pattern that was disrupted only after high flows. A high density of minnows was also associated with a low abundance of algae. Experiments in a pool divided along its length by a fence, and the addition of bass to a pool with many minnows, produced results consistent with the hypothesis that minnow grazing could suppress algae but only where bass were scarce. In pools with bass, the reduced numbers of surviving minnows retreated to the shallow margins, where bass did not follow, while algal biomass increased in the deeper areas. It might be argued that the experiments were not truly replicated (single stream pools were manipulated), but the biological background knowledge was so sound that this drawback was greatly reduced.

This paper was clearly foundational, leading to further contributions by Power and colleagues and, in particular, to Power's highly influential

1990 paper, which revealed a trophic cascade involving direct and indirect interactions among four trophic levels in a California stream in summer (see chapter 10). **Power et al.** (**2008** [chapter 10]) went on to show the context dependence of such a cascade by extending the study over the seasons and changing flow conditions. The occurrence of cascades in ecosystems supported mainly by in situ algal production is now well supported (McIntosh and Townsend 1996; Kohler and Wiley 1997), though cascades may be less frequent in the more diffuse webs of detritus-based ecosystems (Hildrew 2009). More generally, the 1985 paper also showed that species interactions could be quite strong in streams (a somewhat controversial view at the time), contributing to the ongoing debate about the relative roles of abiotic (i.e., disturbance) versus biotic (i.e., species interactions) factors in lotic systems (see chapter 4).

Winemiller, K. O. 1990. "Spatial and Temporal Variation in Tropical Fish Trophic Networks." *Ecological Monographs* 60:331–36.

Winemiller's 1990 study constituted an ambitious and rigorous effort to characterize four river food webs in the lowland tropics of Venezuela. Not only was it unusual in that it was a tropical study, but it dealt with highly complex, speciose communities (the number of fish species varied from twenty to eighty-eight) and sought to count food-web links as well as measure interaction intensities. It produced both annual and seasonal webs, and it related food-web structure to environmental conditions associated with seasonal rainfall. Relative to other studies at the time, this one took a radically different approach. Whereas **Hildrew et al.** (**1985**) attempted to draw a whole food web from a single fishless stream, Winemiller's study, which dealt with diverse tropical food webs, was much more challenging. It was also able to assess for the first time how estimates of some common food-web statistics varied with the level of link intensity adopted (e.g., by eliminating very weak links). As a trade-off against the enormous effort involved, this study dealt only with fish, but also stood out from much of the earlier food-web literature in that it applied consistent methods and sampling effort to the different communities. Food-web links were based on visual inspection of fish gut contents from large samples of specimens. The number of nodes in the webs was increased by dividing gut contents into components such as detritus, aquatic plants, other fish species, terrestrial invertebrates, aquatic insects, and so on. Nevertheless, the level of taxonomic resolution at these lower levels was much less than for the fish.

The food webs for the four ecosystems contained between 58 and 104 "interactive taxonomic units" (fish and non-fish) and between 208 and 1,243 total trophic links (figure 5.2). Food-web diagrams were extremely complex, much more so than webs compiled from the literature (Briand and Cohen 1987). Annual webs (combining all links observed over a year) were inevitably more complex than webs from a single season. Complex as they were, Winemiller wondered "how many more links would have been added to each web if the feeding of molluscs, predacious insects and other organisms (including photosynthetic and heterotrophic microbes) had been counted?" To put this question in context, the most complete version of the Broadstone web achieved, having resolved much of the meiofauna, contained 131 taxa and 841 food-web links, and this for a fishless system (Hildrew 2009).

A major contribution of Winemiller's study was to show just how sensitive food-web statistics are to methods, such as the lumping of species into trophic units, the inclusion of links, and not least, the practical decision to ignore much of the diversity of organisms lower in the food web. In this light, the paper ends with a note of caution on constructing and analyzing food webs, given the variability in methods adopted and in study design. Food webs remain fascinating depictions of nature, however, and this and other subsequent papers have developed the more useful and tractable aspects of food webs—including the strength and nature of species interactions—into an important part of ecology and one in which stream studies have been significant (Hall et al. 2000; Parker and Huryn 2006; Benke 2018; Marcarelli et al. 2020).

Huryn, A. D. 1998. "Ecosystem-Level Evidence for Top-Down and Bottom-Up Control of Production in a Grassland Stream System." *Oecologia* 115:173–83.

Huryn's 1998 study was inspired by a much earlier effort to understand the role of predators in stream food webs. Allen's (1951) production budget for the invasive brown trout (*Salmo trutta*) and its invertebrate prey in the Horokiwi Stream of New Zealand was a classic early attempt to balance a predator-prey production budget. Although the methods

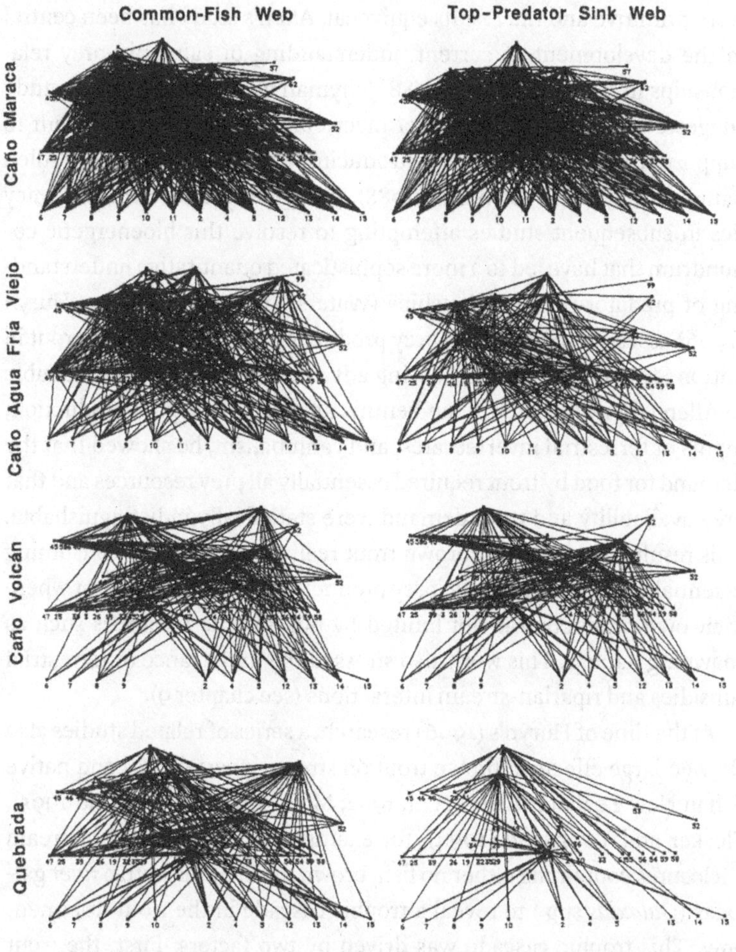


Fig. 5.2 Computer-illustrated community food-web diagrams (*left*) and top-predator sink web diagrams (*right*) for each site, based on combined annual resource matrices. All observed feeding interactions were drawn (LT = 0) as unidirected lines of uniform thickness. (Link threshold, LT, is defined in Winemiller 1990, table 1.) Nodes 2 through 15 represent basal nodes. Horizontal positions of nodes 2 through 62 (nonfish) were fixed based on trophic affiliation with either the terrestrial or aquatic environment, and vertical positions were fixed based on the trophic continuum and rough diet estimates. The vertical position of each node was equal to its trophic level, and horizontal position was calculated as the mean of the horizontal positions of its prey (weighted by their volumetric proportions in the consumer's diet). Numbers specify nodes, which are identified in the appendices to Winemiller (1990). From Winemiller, K. O. 1990. "Spatial and Temporal Variation in Tropical Fish Trophic Networks." *Ecological Monographs* 60:331–36.

were primitive and the results equivocal, Allen's study has been central in the development of current understanding of salmonid-prey relationships in streams (Waters 1988; Jellyman et al. 2000). Allen's study suggested that the production of invertebrate prey was insufficient to support brown trout demand, introducing what is known as the "Allen paradox" (Hynes 1970; Waters 1988), though its main scientific legacy lies in subsequent studies attempting to resolve this bioenergetic conundrum that have led to a more sophisticated quantitative understanding of predator-prey relationships (Waters 1988). For instance, Huryn (1996) assessed the predator-prey production budget for brown trout in Sutton Stream, New Zealand, using advanced approaches not available to Allen (1951). By quantifying benthic and hyporheic prey production, inputs of terrestrial invertebrates, and cannibalism, he showed that the demand for food by trout required essentially all prey resources and that prey availability and trout demand were statistically indistinguishable. This result suggested that brown trout really are capable of consuming essentially all of the available prey production in streams, at least where their own production is not limited by non-trophic resources such as spawning habitat. This work also stressed the importance of terrestrial subsidies and riparian-stream interactions (see chapter 9).

At the time of Huryn's (1996) research, a series of related studies also showed large effects of brown trout on stream invertebrates and native fish in New Zealand (Crowl et al. 1992; McIntosh and Townsend 1995). Flecker and Townsend (1994), for example, used replicated instream enclosures containing either no fish, brown trout, or the native river galaxias (*Galaxias* spp.) to reveal a trophic cascade in the trout treatment only. This trophic cascade was driven by two factors. First, the trout consumed more prey than the river galaxias (reducing grazer density). Second, invertebrates in the presence of trout behaved differently than those in other treatments, remaining hidden beneath shelter during the day and grazing biofilms adjacent to shelter only during the night (McIntosh and Townsend 1994, 1996). Consequently, reduced grazing led to an accumulation of algae in the trout treatments.

Huryn's 1998 study subsequently tested the patterns revealed by these small-scale experiments in his comparison of production budgets in two streams, one dominated by brown trout and another that had retained the native river galaxias. A trophic cascade was apparent only in the brown trout stream. There, the invertebrate prey, subject to strong predatory pressure, consumed only about 21% of annual net primary production, resulting in the accrual of thick algal mats. In contrast, predation pressure on invertebrates was relatively minor in the native river galaxias stream (fish required only about 18% of prey production), and essentially all algal production was consumed. Invasive brown trout are evidently voracious predators in these streams, leading to the "Allen paradox" and producing a trophic cascade. Other cases of extreme predation pressure have since been identified. For instance, bioenergetic data indicate levels of consumption approaching 100% of prey production by predatory invertebrates in a fishless stream in North Carolina (Wallace et al. 1997). Such predatory pressures, though by no means universal, can elicit important responses at the behavioral, community and ecosystem levels of organization.

## LOOKING FORWARD

Since the early contributions of **Cummins et al. (1966)** and **Minshall (1967)**, studies of trophic interactions in streams and rivers have shifted from description to hypothesis testing by the experimental manipulation of resources or thermal and flow regimes (Ledger et al. 2013; Nelson et al. 2020). This shift in approach, plus improvements in taxonomic resolution and methods based on biochemical analyses, has contributed significantly to contemporary understanding of trophic interactions within stream communities (Guo et al. 2016; Venarsky et al. 2018).

What lies ahead for food-web research? Various sages, including the physicist Nils Bohr, have pointed out that "prediction is very difficult, especially when it is about the future," so we should not be too prescriptive. Nevertheless, it seems likely that food-web science will strengthen due to several ongoing developments:

1. An increase in the availability of well-resolved food webs (often produced using molecular methods), providing a powerful database for meta-analytical studies and new insight into food-web dynamics (Brose et al. 2019; Pringle and Hutchinson 2020).
2. An increasingly sophisticated focus on the interconnectedness of meta–food webs at landscape scales (Häussler et al. 2021), particularly in assessing aquatic-riparian linkages (Häder and Barnes 2019; Wu et al. 2022; see chapter 9).

3. An increase in the sophistication of predictive mathematical models of food-web topology and function (Sentis et al. 2021).
4. The development of new methods for identifying components of diet, including fatty acids and amino acids, sometimes combined with analysis of stable isotopes (Guo et al. 2016; Nielsen et al. 2018; Jardine et al. 2020). These methods are developing very quickly and could yet cause a paradigm shift in understanding of the resources supporting stream webs (Hildrew and Giller 2023).
5. The accelerating development of DNA-based methods for analyzing diets, which, once perfected and standardized, are likely to revolutionize food-web studies (Traugott et al. 2021).

Beyond such methodological advances, the focus of much stream research has shifted to "applications"—ranging in scope from diagnosing local problems, such as the effects of a single pollution incident, to broader problems in environmental stewardship and policy, such as the setting of environmental flows. Stream research has found applications in biomonitoring and in the diagnosis of stressors. These applications, however, usually rely on changes in species composition and sometimes on "functional" changes, rather than changes in food-web attributes per se.

Alterations to food webs because of such changes are widespread and profound, however. For instance, in acidified or metal-stressed systems, where fish are often absent, their absence constitutes a switch in the top predators and consequently in food-web function (Hildrew 2018; Hogsden and Harding 2012). Fertilization of a detritus-based stream in North Carolina caused a switch from a predominantly "brown" to a "greener" food web (Bumpers et al. 2017), and a similar switch famously occurred upon fertilization of the Kuparuk River in Arctic Alaska (Peterson et al. 1985). It has often been demonstrated that shifts in land cover and acidity, for instance, have consequences for food webs (e.g., Layer et al. 2013; Collins et al. 2016; Erdozain et al. 2019). Many such results have been based on the use of stable isotopes to infer the reliance of animals on allochthonous and autochthonous resources (pioneered in stream ecology by Winterbourn and colleagues; Rounick et al. 1982). We are not aware, however, of examples where changes in food-web structure or function have been proposed explicitly to diagnose environmental stress.

In terms of anthropogenic environmental perturbations, however, the "elephant in the room" is the effects of climate change. These effects are often accompanied by shifts in the geographic ranges of stream organisms, either by introductions or by natural changes in response to the climate (Ledger et al. 2013; O'Gorman et al. 2019; Sentis et al. 2021). Consequently, freshwater food webs are subject to increasing instability and "rewiring" (Bartley et al. 2019). Although studies of food webs might not aid in the general alleviation of the negative consequences of climate change, it is important that ecologists document the long-term effects of this "global human experiment" on food-web architecture and dynamics.

# 6

# LITTER DECOMPOSITION

## SHINING A LIGHT ON EARLY INSIGHTS INTO A KEY ECOSYSTEM PROCESS IN STREAMS

Mark O. Gessner, Luz Boyero, and Jennifer L. Tank

The recognition of an intimate connection between streams and their catchments has helped to shape the conceptual foundations of stream ecology. H. B. N. Hynes's cornerstone paper "The Stream and Its Valley" (**Hynes 1975** [chapter 2]) epitomized this view of stream ecosystems as embedded in the landscape, and the linkage between stream channels and riparian vegetation highlighted by K. W. Cummins (1974) became a central building block of the river continuum concept (**Vannote et al. 1980** [chapter 3]), the single most important theoretical framework in stream ecology to date. Central to this perspective is the pivotal role of allochthonous organic matter, most notably in the form of coarse particulate organic matter (CPOM; >1 mm in diameter) such as leaf litter and wood derived from riparian plants.

The importance of plant litter in ecosystems was recognized much earlier, however, and can be traced back at least to Darwin (1883), who devoted an entire book to the topic. For streams, it was Thienemann (1912) who emphasized the critical role of leaf litter as a resource for food webs and carbon cycling early on. Yet it took another fifty years, and inspiration by the pioneering conceptual work of Lindeman (1942), before organic matter and its importance for benthic fauna (see chapters 5 and 9) gained center stage in freshwater ecology. This new emphasis initiated an innovative chapter in stream research. Accordingly, the traditional focus on the responses of organisms to abiotic and biotic conditions shifted toward their role as drivers of ecosystem processes, or what we now collectively call ecosystem functioning.

Much attention has also been given to dissolved organic matter (DOM), some of which is received from surrounding land and some released instream (Battin et al. 2008). In addition, the importance of fine particulate organic matter (FPOM; <1 mm in diameter) has been clearly recognized (Wotton 1994), although the dynamics of this large stock of organic particles remains far less well investigated than the dynamics of CPOM and DOM. It is FPOM that had traditionally been referred to as "detritus," a term that was later extended to include plant litter (CPOM) and DOM (Wetzel 1995). Despite important differences in their size, quality, and dynamics, all these forms of organic matter share a broadly similar fate once supplied to stream channels. They are either (1) transported downstream, where the deposition of particulate organic matter in lakes and oceans potentially leads to its long-term sequestration; (2) transformed to other forms of organic matter differing in size and composition, including microbial and consumer biomass; or (3) mineralized to carbon dioxide and other inorganic compounds (**Petersen and Cummins 1974**; Gessner et al. 1999; Marks 2019). Thus, both particulate and dissolved forms of organic matter are instrumental in governing ecosystem metabolism and elemental cycles in streams.

This chapter focuses on the decomposition of leaf litter, notwithstanding the importance of other forms of organic matter for stream ecosystem functioning and the differences among these forms in their decomposition dynamics. For example, macromolecular DOM, once cleaved by extracellular microbial enzymes, can be assimilated by microorganisms, but not by invertebrates, and wood submerged in freshwater decomposes only at the surface, leading to extremely slow decomposition of large logs and correspondingly long residence times in streams (e.g., Boyd et al. 2012). We highlight six foundational papers that have paved the way for understanding decomposition dynamics, its critical drivers, and the key organisms involved in the process. These papers are not ordered chronologically, but follow the logic of starting with information on litter mass loss, followed by insights into the role of fungal decomposers in relation to litter quality and dissolved nutrient availability, and finishing with the role of detritivorous macroinvertebrates in the decomposition process, both at the local and, finally, the global scale. These six papers were chosen from a sizable body of original literature published more than twenty-five years ago, while acknowledging that many other influential studies have appeared both since and before that time.

## HIGHLIGHTED PAPERS

Kaushik, N. K., and H. B. N. Hynes. 1971. "The Fate of the Dead Leaves That Fall into Streams." *Archiv für Hydrobiologie* 68:465-515.

How could we start off the list of foundational papers in this chapter with a publication other than Kaushik and Hynes 1971, which laid the groundwork for an entirely new field of inquiry in stream ecology? Their study, already recognized as a Citation Classic four decades ago (Kaushik 1982), has inspired several generations of stream ecologists seeking a better understanding of the litter decomposition process. As Narinder Kaushik recounts in his 1982 paper, the impetus for the study had been a growing interest in the early 1960s (e.g., **Ross 1963** [chapter 1]) in understanding the food base of stream invertebrates and a recognition of organic matter derived from riparian vegetation as a critical resource.

Despite their initial motivation to explore the food source of stream invertebrates, Kaushik and Hynes also began to examine the decomposition process itself. To this end, they deployed leaf litter in two streams to determine decomposition rates under field conditions. In addition, they quantified the immobilization of nitrogen (N) by decomposing leaves, suspecting that the stoichiometric mismatch, as we would call it today, between high-C:N food and low-C:N consumer biomass would drive the rate and efficiency of resource use. As a complement to this field experiment—the second field experiment on litter decomposition in streams after that of Matthews and Kowalczewski (1969)—Kaushik and Hynes designed a suite of laboratory studies intended to shed light on the drivers of decomposition and on the mechanistic underpinnings of variation in decomposition rates and N immobilization. These experiments led to the conclusions that decomposition rates depend on litter quality, and hence on the composition of riparian vegetation; on water quality, particularly dissolved nutrient content; and on temperature. Moreover, the study showed that fungi rapidly colonize leaf litter in streams and contribute to its decomposition, not only by direct degradation of plant tissue, but also by enhancing the palatability and nutritional quality of leaves for litter-consuming stream detritivores (figure 6.1), thereby promoting invertebrate feeding—a mechanism that was later called conditioning (Cummins 1974; **Petersen and Cummins 1974**).

Kaushik and Hynes's study, together with the pioneering study by **Petersen and Cummins (1974)**, sparked immense interest, triggering

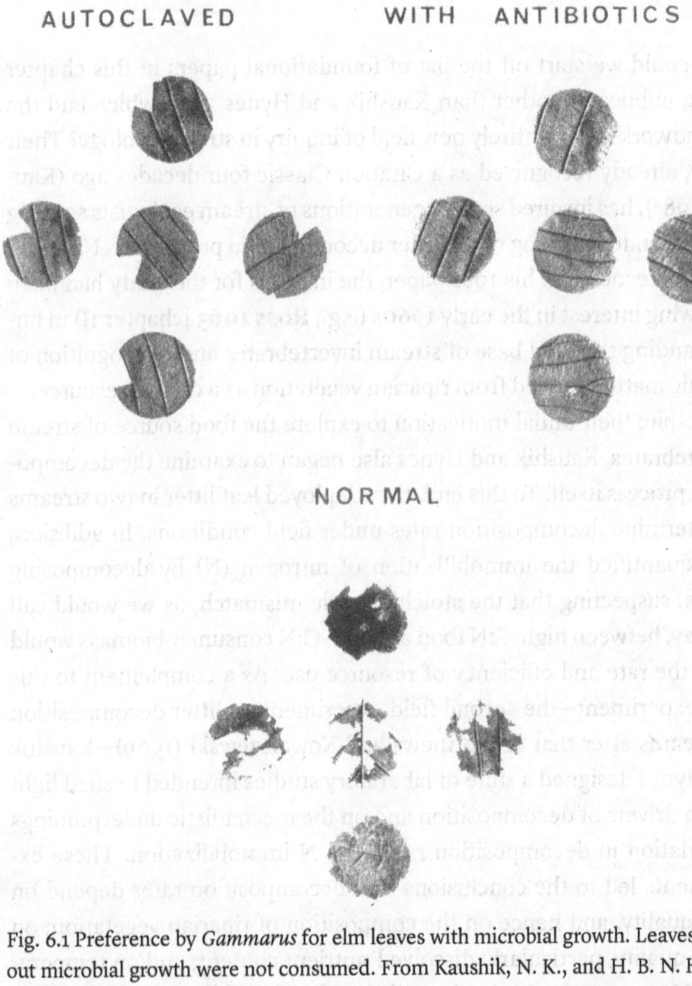


Fig. 6.1 Preference by *Gammarus* for elm leaves with microbial growth. Leaves without microbial growth were not consumed. From Kaushik, N. K., and H. B. N. Hynes. 1971. "The Fate of the Dead Leaves That Fall into Streams." *Archiv für Hydrobiologie* 68:465–515.

an avalanche of investigations on litter decomposition, the association of macroinvertebrates using litter as food and habitat, trophic relationships between litter-colonizing fungi and detritivores, and other related topics. Immediate consequences of that interest included follow-up studies in which, in a historically remarkable parallelism, stream and microbial ecologists joined forces in two distinct laboratories to explore the role of fungi in stream ecosystem functioning. These initiatives resulted in groundbreaking discoveries, with the early landmark studies characterizing the fungal communities in decomposing leaves (Bärlocher and Kendrick 1974; Suberkropp and Klug 1976) and elucidating trophic interactions between leaf-colonizing fungi and stream detritivores (Bärlocher and Kendrick 1973).

Petersen, R. C., and K. W. Cummins. 1974. "Leaf Processing in a Woodland Stream." *Freshwater Biology* 4:343–68.

In the 1970s, more than forty years after the first leaf-litter decomposition studies had been conducted in forests, the time was also ripe for investigating the process in forest streams. This was underscored especially by a comprehensive paper by Petersen and Cummins that presented a wealth of data embedded in a compelling conceptual framework. Reinforced by Cummins's (1974) broader, visionary theoretical underpinning, the paper became a Citation Classic (Petersen and Cummins 1991) even before the second wave of studies on leaf-litter decomposition began in the early 1990s. Attesting to its tremendous influence, it is today the most cited original study on litter decomposition in streams, with only Webster and Benfield's (1986) landmark synthesis on vascular plant decomposition in freshwaters being more frequently cited. This exceptional impact is an ironic twist of fate, given that the paper was originally rejected by the leading ecological journal at the time (R. C. Petersen, personal communication).

Grounded in fundamental questions about the energetic basis of streams (see chapter 7), Petersen and Cummins's study quantified leaf decomposition rates for fifteen species of deciduous trees at multiple sites along a woodland stream in autumn and winter. The results of this field study, which involved more than a thousand standardized leaf packs, underscored the role of litter quality and temperature as dual drivers controlling decomposition rates. Recognizing the food-web and ecosystem-level implications of the large differences among tree species in leaf-litter decomposition rates, Petersen and Cummins framed their results as a "processing continuum" that extends the annual autumn pulse of allochthonous litter as a resource for stream microbes and macroinvertebrates over many months (Cummins 1974; figure 6.2), reinforcing the notion that leaf litter forms the key basis of food webs in streams draining temperate deciduous forests (Wallace et al. 1997). In addition, Petersen and Cummins introduced the exponential decay model to express the mass loss of submerged leaves over time, resulting in a metric for leaf-litter decomposition that was comparable to those used in earlier studies in terrestrial environments. The rate coefficient of this model (*k*, expressed per day or per degree-day) has since become the standard metric to compare decomposition rates among stream locations, plant species, or experimental treatments.

In contrast to the commonly used mesh bags, Petersen and Cummins championed a leaf-pack technique that mimicked unconfined litter trapped by retention structures in stream channels. Although the

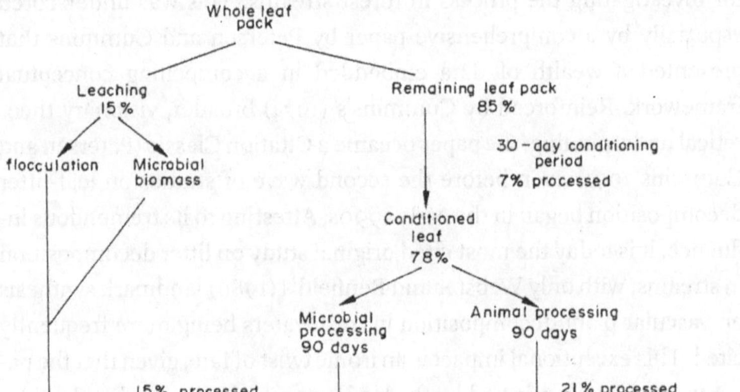


Fig. 6.2 Summarized processing budget for leaf-pack material. From Petersen, R. C., and K. W. Cummins. 1974. "Leaf Processing in a Woodland Stream." *Freshwater Biology* 4:343–68.

unconfined leaves better reflected the decomposition rates of natural leaf packs than did leaves enclosed in bags (Cummins et al. 1980), the leaf-pack approach was not widely adopted, possibly due to concerns about the loss of large leaf fragments. Nevertheless, combined with the calculation of exponential decay coefficients, Petersen and Cummins's 5 g litter batches became the standard experimental approach to address a broad range of questions relating to litter decomposition in streams.

Gessner, M. O., and E. Chauvet. 1994. "Importance of Stream Microfungi in Controlling Breakdown Rates of Leaf Litter." *Ecology* 75:1807–17.

Gessner and Chauvet provided a new quantitative perspective on the contribution of fungi to leaf-litter decomposition in streams, building on pioneering work by Bärlocher and Kendrick (1974) and Suberkropp and Klug (1976), in addition to that of **Kaushik and Hynes (1971)** and **Petersen and Cummins (1974)**. This achievement was possible thanks to a newly developed method (Newell et al. 1988; Gessner et al. 1991) of estimating fungal biomass in decomposing litter. Two lines of evidence suggested to Gessner and Chauvet that fungi play a key role in the decomposition process. First, estimates of fungal biomass were surprisingly high, ranging from 6.5% to 18.4% of total leaf mass. Second, variation in decomposition rates was strongly related to differences in fungal biomass accrual among different litter types. Similarly strong negative relationships emerged with the lignin concentration of the litter, which matched findings in forests, where decomposition rates were negatively related to the lignin:N ratio of litter (Melillo et al. 1982). These observations suggested that decomposition rates in streams are controlled by litter carbon quality, and that this control is mediated by fungi, whether due to their direct degradative activity, their stimulation of detritivore feeding, or both.

Another quantitative approach to assessing the performance of fungi on decomposing leaves in streams had already been used more than a decade earlier (Bärlocher 1982). The method takes advantage of the capability of aquatic hyphomycetes, the main fungi colonizing and degrading leaves in streams (Bärlocher 1992), to sporulate abundantly underwater, with the number of spores released from submerged leaves reflecting fungal reproduction rates. Using a similar method, Gessner and Chauvet found that the maximum rates of sporulation on different litter types over the course of decomposition correlated with both litter lignin concentration and decomposition rate, as had maximum fungal biomass, reinforcing the conclusion that fungi play a mediating role in the control of decomposition by litter carbon quality.

This finding, which was corroborated in subsequent studies, paved the way for including measurements of fungal biomass as a routine response variable in a wide range of investigations on litter decomposition in streams. These investigations included studies aimed at improving understanding of the mechanisms and controls of decomposition (Gulis and Suberkropp 2003a; Jabiol et al. 2019), assessments of the relative importance of fungi versus bacteria (Pascoal and Cássio 2004), and the use of litter decomposition and associated processes in functional stream bioassessment (Chauvet et al. 2016) and ecotoxicology (Fernández et al. 2015).

Suberkropp, K., and E. Chauvet. 1995. "Regulation of Leaf Breakdown by Fungi in Streams: Influences of Water Chemistry." *Ecology* 76:1433–45.

Understanding the role of external drivers in controlling litter decomposition rates has been a common objective in studies of stream organic matter dynamics. In particular, dissolved nutrient concentrations had been identified as a key factor in the 1970s, in addition to temperature, streamflow and turbulence, dissolved oxygen, and other environmental variables (Webster and Benfield 1986). One breakthrough in this line of inquiry came from Suberkropp and Chauvet, who used a natural gradient across eight forest streams to document a strong relationship between dissolved nutrient concentrations and not only leaf mass loss, but also microbial biomass (measured as adenosine triphosphate; Abelho 2020) and fungal sporulation rate. In addition, a reciprocal transplant experiment showed that the microbial response to increased nutrient availability was rapid, and that the imprint of previous exposure to elevated nutrient supply was long lasting. Combined with data from earlier field and laboratory studies, these results provided compelling evidence that nutrients act as a key environmental driver of leaf-litter decomposition in streams by controlling microbial activity.

These relationships offered a mechanistic explanation for the results of earlier nutrient-enrichment studies conducted in experimental stream channels (Triska and Sedell 1976) and natural streams (Elwood et al. 1981; Newbold, Elwood, Schulze et al. 1983). Elwood et al.'s (1981) whole-stream nutrient addition, in particular, clearly demonstrated the stimulating effect of dissolved phosphorus on microbial respiration and leaf mass loss. A companion experiment involving ammonium enrichment failed to detect a similar stimulating effect (Newbold, Elwood, Schulze et al. 1983), which was consistent with the notion that leaf-associated microbial activity was limited by phosphorus availability in the study stream. This conclusion was further supported by a radiotracer experiment designed to test nutrient-spiraling theory, in which quantification of the immobilization of experimentally added $^{32}$P-phosphate by various stream compartments showed that decomposing litter accounted for the largest fraction (60%) of the total $^{32}$P uptake (**Newbold, Elwood, O'Neill et al. 1983** [chapter 8]).

Numerous follow-up studies have since tested to what extent dissolved nutrient supply (e.g., Gulis and Suberkropp 2003a; Ferreira et al. 2006; Woodward et al. 2012) and stoichiometry (Rosemond et al. 2015) enhance litter-associated microbial activity and increase leaf-litter decomposition rates. Two comprehensive meta-analyses concluded that nutrient effects on litter decomposition are common (Ferreira et al. 2015; Ardón et al. 2021). However, the strength of these effects depends on (1) background concentrations in stream water (Ferreira et al. 2015), with saturation reached at relatively low concentrations (Ferreira et al. 2006); (2) the interaction between nutrient supply in stream water and litter (Stelzer et al. 2003); and (3) litter recalcitrance, because slow carbon acquisition by microbial decomposers reduces nutrient demand (Jabiol et al. 2019). It is now also firmly established that the influence of nutrients on decomposition is not limited to leaves. Decomposition rates of wood, associated fungal activity and biomass, and microbial respiration all increase with elevated nutrient availability (Tank and Webster 1998; Ferreira et al. 2006); wood decomposes very slowly in freshwaters due to the combination of inherently high carbon-to-nutrient ratios, poor carbon quality, and the fact that microbial decomposition is largely limited to wood surfaces. Consequently, microbial biofilms on wood will likely compete for dissolved nutrients with other stream ecosystem compartments such as decomposing leaf litter or periphyton, especially when nutrient concentrations in stream water are low (Tank et al. 1998).

Wallace, J. B., J. R. Webster, and T. F. Cuffney. 1982. "Stream Detritus Dynamics: Regulation by Invertebrate Consumers." *Oecologia* 53:197–200.

Shredders are a functional feeding group of benthic macroinvertebrates whose mouthparts enable them to ingest leaves and other types of CPOM efficiently (Wallace and Webster 1996). Wallace, Webster, and Cuffney were the first to demonstrate at the whole-stream scale how the presence of these detritivores influences litter decomposition, supporting earlier small-scale studies showing the consumption capacity of shredders feeding on leaf litter (Cummins 1973). The long-term whole-ecosystem experiment by Wallace and colleagues consisted of applying an insecticide to one of two small, forested streams and measuring mass loss of four types of leaf litter differing in recalcitrance, as well as changes in the standing crop of benthic organic matter and export of suspended FPOM at the whole-stream scale. The insecticide treatment drastically reduced shredder numbers and slowed decomposition, although microbial decomposers were not affected, as indicated by the respiration rates of treated and untreated leaf discs and by direct observations of aquatic hyphomycetes. In addition, the insecticide treatment increased the standing stock of benthic organic matter and substantially reduced FPOM export downstream. Furthermore, shredder biomass and decomposition rates recovered within two years after discontinuing the insecticide treatment. A follow-up insecticide application, swapping the treatment and reference streams, with additional controls included in the experimental design, produced essentially the same results (Cuffney et al. 1990). Collectively, these data provided strong evidence that the observed insecticide effects on litter decomposition and on ecosystem-level organic matter processing and export were due to the removal of shredders.

The notion that shredders play a key role in the decomposition of leaf litter and the incorporation of litter carbon and nutrients into stream food webs is now firmly established (Marks 2019). The application of insecticides, or other toxic substances, to whole ecosystems is no longer an option, however, even in experimental catchments. The ethical concerns may be the chief reason why no other whole-stream manipulations have repeated these pioneering experiments to assess the role of shredders in litter decomposition at the ecosystem level. This study, and that of Cuffney and colleagues (1990), are thus to be valued for their one-of-a-kind contributions. Nevertheless, González and Graça (2003) successfully combined laboratory and field observations to quantify the consumption of litter by a shredding caddisfly at the whole-stream scale.

Hundreds of other studies have addressed the question of the extent to which shredders colonize leaf litter and contribute to its decomposition (e.g., Richardson 1992), and many of them have corroborated the pioneering results of Wallace and colleagues showing the major role of shredders in decomposition. These later studies have confirmed the critical contribution of shredders through estimates of consumption rates and shredder biomass in litter bags (Hieber and Gessner 2002), and by the use of underwater electric fencing to prevent shredders from colonizing batches of leaves (Moulton et al. 2019). Most of these studies, however, have used paired litter bags (Gessner and Chauvet 2002), comparing decomposition rates of litter enclosed in coarse-mesh bags (to which shredders have access) with those of litter in fine-mesh bags (where decomposition is primarily microbial), which enables partitioning of the contribution of shredders to decomposition (Lecerf 2017). Although this method has several drawbacks (e.g., influence of mesh size on mechanical fragmentation, exclusion of macroconsumers such as crabs or crayfish; Moulton et al. 2019), it has generally supported Wallace and colleagues' conclusion about the critical role that shredders play in the decomposition of leaf litter in forest streams.

Irons, J. G., III, M. W. Oswood, R. J. Stout, and C. M. Pringle. 1994. "Latitudinal Patterns in Leaf Litter Breakdown: Is Temperature Really Important?" *Freshwater Biology* 32:401–11.

Irons, Oswood, Stout, and Pringle set the stage early on for analyses of broad-scale patterns of litter decomposition in streams by conducting the first litter breakdown experiment in streams over a wide geographic range spanning a climatic gradient. Specifically, their 1994 study compared the decomposition rates of ten litter types at three locations along a 55° latitudinal gradient from Costa Rica to Michigan to Alaska. Decomposition was much faster in the tropics than at the other two sites when rates were expressed per day, but rates were higher in the boreal stream when normalized for temperature (i.e., expressed per degree-day). While the study included just three locations, and had no replication within regions, the observed pattern hinted for the first time that decomposition rates vary systematically across climatic zones. Using data from the literature to complement the results of their own experiment, Irons and colleagues attributed this variation to differences in the temperature-dependent contribution of microbial and shredder communities to decomposition.

The authors' focus on exploring broad-scale patterns arose because decomposition studies had been mostly restricted to small geographic areas, commonly one or several streams within a single catchment or region in North America or Europe. Coverage of the tropics was poor, although some of the earlier investigations were carried out in streams near the equator (Stout 1980; Dudgeon 1982). This geographic bias and narrow scope sparked the question of how representative the locally observed patterns are at continental and global scales, with a particular emphasis on variation in the tropics (Boyero et al. 2009). Irons and colleagues set this train of inquiry in motion, although strong momentum was gained only about a decade ago (e.g., Boyero et al. 2011; Woodward et al. 2012; Tiegs et al. 2019; Boyero, Gessner et al. 2021).

The relationships found in this study were later substantiated by more comprehensive analyses that took advantage of meta-analyses (Follstad Shah et al. 2017) and large-scale coordinated experiments at multiple locations around the world (Boyero et al. 2011). Overall, it has become increasingly evident that leaf-litter decomposition is highly dependent on multiple factors that vary across climatic gradients (Graça et al. 2015; Boyero, López-Rojo et al. 2021; Boyero, Pérez et al. 2021). In particular, globally coordinated distributed experiments have shown that rates of microbial decomposition increase toward the equator (Tiegs et al. 2019) and that shredders tend to contribute a greater share to decomposition toward higher latitudes (Boyero et al. 2011)—patterns reported initially by Irons and colleagues across their three study sites. Temperature is now recognized as a critical driver behind these patterns (Follstad Shah et al. 2017), as Irons and colleagues had already suggested. However, we now know that latitudinal variation in the structure of riparian vegetation (Boyero, Pérez et al. 2021; Kominoski et al. 2013) and the geographic distribution of shredders (Boyero, López-Rojo et al. 2021) and fungal decomposer communities (Seena et al. 2019) are also important factors influencing the broad-scale patterns.

## LOOKING FORWARD

Our understanding of litter decomposition in streams has come a long way since the pioneering work by Kaushik, Hynes, Petersen, and Cummins, and it has formed a critical foundation for highlighting the importance of riparian linkages in shaping stream communities and ecosystem functioning (**Gregory et al. 1991** [chapter 9]). Well over a thousand original investigations of litter decomposition in streams and rivers have since been published, many of them in recent years. This wealth of information has been summarized in a recent book devoted entirely to litter decomposition in streams (Swan et al. 2021), complemented by a second book that compiles methods for studying the decomposition process (Bärlocher et al. 2020). Moreover, detailed data on litter decomposition in streams have been synthesized in meta-analyses (Ferreira et al. 2015; Ardón et al. 2021) and in more than a dozen reviews offering a wide variety of perspectives. These reviews have covered methodological considerations (Boulton and Boon 1991; Marks 2019); the roles of fungi (Gessner et al. 2007; Gulis et al. 2019) and macroinvertebrates (Anderson and Sedell 1979; Graça 2001); the importance of litter, decomposer, and consumer diversity (Gessner et al. 2010); conceptual considerations (Gessner et al. 1999; Graça et al. 2015); and comprehensive treatments of the topic (Bird and Kaushik 1981; Webster and Benfield 1986; Abelho 2001; Tank et al. 2010; and Marks 2019).

The range of perspectives has greatly expanded since the early 1970s, while the attention received by different aspects of litter decomposition has varied considerably over time. Some issues of central interest in recent years had escaped imagination at the time, and others are emerging only now. For example, interest in the extent to which the diversity of riparian plants, fungal decomposers, and leaf-shredding macroinvertebrates influences decomposition has had a strong and consistent momentum over the past twenty years (Gessner et al. 2010; Swan et al. 2021). Likewise, assessments of impacts of various aspects of global change have highlighted the sensitivity of the litter decomposition process in streams to changing environmental conditions (e.g., Ferreira et al. 2015; Follstad Shah et al. 2017; Kominoski et al. 2013; Swan et al. 2021). Finally, metrics associated with litter decomposition have been considered in stream ecosystem assessment (Gessner and Chauvet 2002) and found useful as functional endpoints in ecotoxicology (Fernández et al. 2015; Swan et al. 2021).

At the same time, some of the initial questions remain unresolved. For example, although it has been firmly established that shredders exhibit high selectivity for fungi, being capable of distinguishing among individual fungal species (Arsuffi and Suberkropp 1985), it remains unclear which litter properties are responsible for triggering preferential feeding. In addition, while great strides have been made in understanding the drivers of overall litter mass loss, very little quantitative information is available about the fate of the leaf material that is being decomposed (Baldy et al. 2007; Marks 2019). Interactions between fungi and bacteria in decomposing litter, and their repercussions for the decomposition process (Gulis and Suberkropp 2003b), are virtually unknown, perhaps due to the effort required to quantify microbial performance in leaf litter. These are just three fundamental issues that await clarification.

A broad range of new and unexploited analytical methodologies have potential to shine a light on some of these unresolved issues. Approaches that will be applied more commonly in the future range from the suite of -omics and isotopic tracer techniques, including quantitative stable-isotope probing, to the automation of fungal and shredder community analyses by automated image acquisition and processing using machine learning. Some of these methods are likely to produce data on litter decomposition at a completely new level of resolution and, combined with evolving concepts and models, promise to broaden and deepen understanding of the litter decomposition process. These studies will be the basis of the future foundational papers on litter decomposition in streams.

# 7

# STREAM ECOSYSTEM ENERGETICS

Robert O. Hall Jr. and Emily S. Bernhardt

Ecosystem energetics comprises all biological processes in an ecosystem that contribute to the fixing of inorganic to organic carbon by autotrophic organisms and the breakdown of organic carbon back to its mineral $CO_2$ form. This idea is not only a cornerstone of ecology, but is of considerable modern interest given that the world's biota drive the largest annual fluxes in and out of the atmospheric $CO_2$ pool. Streams and rivers contribute strongly to this global $CO_2$ budget with net emissions that are approximately equal to global terrestrial net ecosystem production (Raymond et al. 2013; Battin et al. 2023).

The study of ecosystem energetics was in many ways born in streams. H. T. Odum adapted Sargent and Austin's (1949) method of budgeting oxygen to estimate photosynthesis (gross primary production, GPP) and respiration (ecosystem respiration, ER) in coral reefs and applied it to well-mixed flowing waters. This method allowed Odum to estimate GPP and ER at the stream reach scale (**Odum 1956**). The balance of these fluxes is net ecosystem production (NEP; Woodwell and Whittaker 1968). Ecosystems where GPP > ER are autotrophic, and either store or export organic matter (Lovett et al. 2006), while ecosystems where GPP < ER are heterotrophic, and rely on subsidies of organic matter to support this excess respiration. The Odum approach has a terrestrial analog: ecologists estimate terrestrial ecosystem metabolism using eddy covariance (Baldocchi 2020). This method estimates net ecosystem exchange as the vertical flux of $CO_2$ extracted from turbulent air flow over the canopy-atmosphere interface. Despite their vastly different methods and mathematics, Odum's methods and eddy covariance are conceptually similar, but that similarity is not well appreciated in the literature, and **Odum (1956)** is not cited in the foundational papers on eddy flux techniques.

Stream ecologists apply the ecosystem concept to streams by treating a stream reach as an entity with defined boundaries across which inputs and outputs of energy and materials are measured (Hasler 1947; Bormann and Likens 1967). This idea enabled accurate organic matter budgets that reinforced Odum's theory that streams are mostly heterotrophic ecosystems. In fact, early organic matter budgets showed that a stream's energetic basis can depend nearly exclusively on organic matter inputs received from the surrounding terrestrial ecosystem (**Fisher and Likens 1973**). This role of allochthonous inputs in driving metabolism in rivers became a cornerstone of stream ecology, and indeed, streams became the best example of subsidized ecosystems (Polis et al. 1997). The energetic dominance of heterotrophy and terrestrial inputs was engrained to such a degree that stream ecologists needed reminding that algal production can be the primary source of energy and food for freshwater food webs, even in highly heterotrophic ecosystems (**Minshall 1978**).

The river continuum concept (RCC; **Vannote et al. 1980** [chapter 3]) explicitly predicted shifts in ecosystem energetics from headwaters to the largest rivers. The RCC predicted that productivity would be maximized in unshaded, clear river segments and minimized in heavily shaded streams or turbid or blackwater rivers. Respiration rates would follow the opposite trajectory, being maximized where terrestrial subsidies are highest, typically in headwaters, where high edge-to-volume ratios result in large areal inputs of terrestrial organic matter (**Vannote et al. 1980**). **Meyer and Edwards (1990)** expanded the ideas of the RCC to include the role of the floodplain in driving high respiration rates in large blackwater rivers, while **Fellows et al. (2001)** further expanded the boundaries to incorporate the important ecosystem contribution of respiration from the hyporheic zone, that region of the stream subsurface where surface water and groundwater are exchanged.

One of the primary challenges of applying the ecosystem concept to rivers is that the frequent removal and replacement of biomass due to floods and droughts often violates the assumption of a steady state.

Thus, estimates of productivity based on biomass accumulation, commonly used in terrestrial and lake ecosystems, are not reliable. Similarly, single-day estimates of metabolism from one or a few days are unlikely to represent annual productivity in hydrologically dynamic rivers (Bernhardt et al. 2018). **Hall (1972)** first documented seasonal variation in ecosystem metabolism, demonstrating that a forested stream has the highest fluxes of both GPP and ER in the spring and autumn, with both rates suppressed during the summer due to canopy shading. **Uehlinger (2000)** applied this approach at much higher temporal resolution to demonstrate the high day-to-day variation in metabolic rates due to floods and to quantify ecosystem resistance and resilience to these disturbances. In this chapter, we highlight six foundational papers, beginning with **Odum's (1956)**, that developed and expanded the study of riverine energetics.

## HIGHLIGHTED PAPERS

Odum, H. T. 1956. "Primary Production in Flowing Waters." *Limnology and Oceanography* 1:102–17.

Howard T. Odum's 1956 paper was a breakthrough in the field of ecosystem ecology at large. This paper did two things simultaneously. First, it proposed a method to estimate metabolism (GPP and ER) in a Florida spring-fed stream by measuring dissolved oxygen (DO) throughout a day. GPP increases DO during the day, while ER lowers the DO concentration below saturation at night—all while accounting for the flux of gas at the air-water interface. The beauty of this method is that it allows for estimates of metabolism at the scale of a long reach of stream, effectively smoothing over the smaller-scale variability that one needs to contend with when using experimental chambers. Thus, one can derive an estimate of metabolism at the ecosystem level. Spring-fed streams were useful places to test this approach because they are stable and highly productive, allowing for large diel variation of DO. Odum's idea of using this method for a stream came from studies of coral reefs, when he and his brother Eugene estimated metabolism at Eniwetok Atoll (Odum and Odum 1955). The Odums got the idea from oceanographers Sargent and Austin—one of several instances in which marine ecology has advanced stream ecology. Sixty-eight years later, we are still using the same basic model to estimate metabolism in streams and rivers. Improvements have come from better estimates of gas exchange (Marzolf et al. 1994; Young and Huryn 1998) and inverse statistical methods that allow for directly estimating rates of gas exchange from the DO data themselves as well as accounting for uncertainty in the estimates (e.g., Holtgrieve et al. 2010; Appling, Hall et al. 2018). We note that the idea of estimating gas exchange from DO data is itself derived from Odum, who, in his 1956 paper, outlined the general approach of measuring the rate of decline of DO following sunset as a function of its saturation deficit. Hornberger and Kelly (1975) later formalized this approach as the nighttime regression method.

The second contribution of this paper was to the theory of energetics in ecosystems. Odum conceptualized how ecosystems fall along a gradient of autotrophy versus heterotrophy (figure 7.1). Ecosystems with high amounts of ER relative to GPP must receive external subsidies of organic matter from some other ecosystem, and this idea is central to early (and current) stream ecosystem theory (**Fisher and Likens 1973**; **Wallace et al. 2015** [chapter 10]). We still use this same GPP versus ER conceptualization today as a way of understanding variations in metabolism, both among ecosystems (Hoellein et al. 2013; Bernhardt et al. 2022) and over time in any one ecosystem (Bernhardt et al. 2018; Arroita et al. 2019).

Hall, C. A. S. 1972. "Migration and Metabolism in a Temperate Stream Ecosystem." *Ecology* 53:585–604.

In his 1972 paper, Charles Hall (no relation to Robert O. Hall Jr.) applied the methods developed by **Odum (1956)** to address questions about the energetic and biogeochemical consequences of fish migration in the temperate river network of New Hope Creek, a forested stream in the North Carolina Piedmont, USA. This is a foundational paper for ecosystem ecology because in it, Hall established the first documented ecosystem phenology from diel patterns of gas exchange. This work in a small stream was done well ahead of efforts in terrestrial ecosystems that applied eddy covariance methods to measure daily rates of GPP and ER. Using the methods of Odum and applying them over the course of several years, Hall documented strong seasonality in energy availability in this forested stream that was entirely distinct from, yet

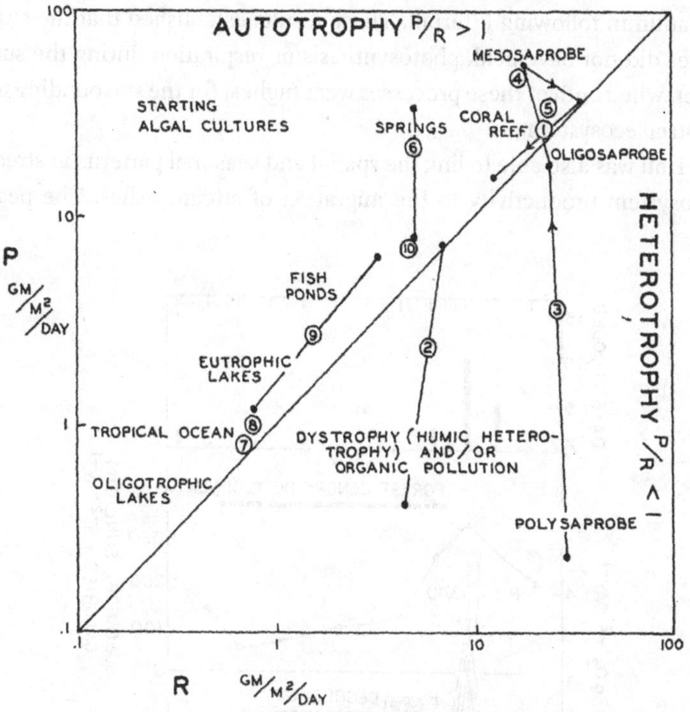


Fig. 7.1 A diagram for the functional classification of communities according to the total metabolism and relative dominance of autotrophic and heterotrophic metabolism. Specific examples are indicated as follows [figure and table numbers refer to Odum 1956]: (1) coral reefs (Table 2); (2) Itchen River, England (Figure 4 and Table 2); (3) recovery sequence on the White River, Indiana (Figure 5); (4) Birs, Switzerland (Table 2); (5) turtle grass (Table 2); (6) Silver Springs (Table 2); (7) tropical ocean (after Riley 1953); (8) Linsley Pond, Connecticut (Riley 1940); (9) Little John Pond, Wisconsin (Juday, Blair, and Wilda 1943); (10) fish ponds (from Winberg 1937, as quoted by Juday et al. 1943). Ryther (1954) described the behavior of the *P:R* ratio in algal cultures. From Odum, H. T. 1956. "Primary Production in Flowing Waters." *Limnology and Oceanography* 1:102–17.

dependent on, the vegetation in its surrounding floodplain (figure 7.2). Indeed, Hynes featured this figure in his famous paper, "A Stream and Its Valley" (**Hynes 1975** [chapter 2]). Hall documented a large peak in instream photosynthesis and microbial activity in spring that was initiated by rising temperatures and subsequently reduced due to a closing forest canopy. Hall also documented a second peak in microbial activity in autumn following litterfall. These results established that the study river did not have peak photosynthesis or respiration during the summer, when both of these processes were highest for the surrounding terrestrial ecosystem.

Hall was also able to link the spatial and seasonal patterns in stream ecosystem productivity to the migration of stream fishes. The peaks

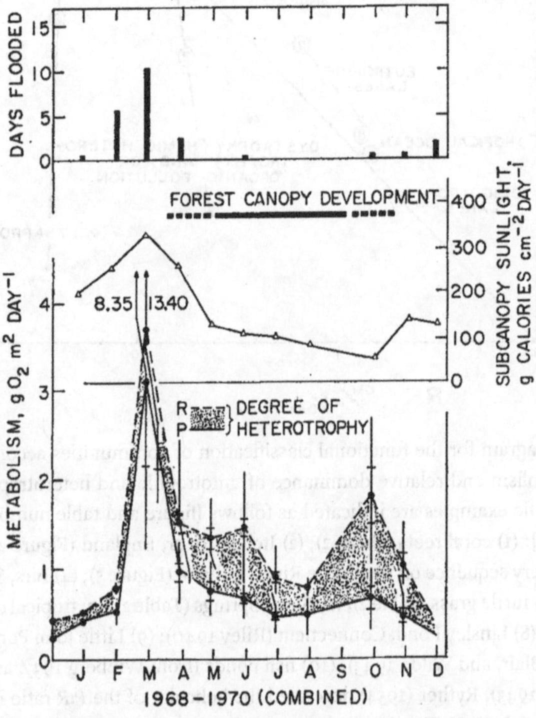


Fig. 7.2 Annual areal oxygen metabolism and some associated phenomena at station 6. *Top*: Mean number of days per month of heavy floods, defined as at least 76 cm above the level of zero streamflow. *Middle*: Mean monthly subcanopy insolation determined on days on which stream metabolism was also measured. Included is an approximate representation of the forest-canopy shading effect. *Bottom*: Annual oxygen gross photosynthesis (P) and gross community respiration (R) based on sixty diurnal analyses over twenty-seven months. Points represent mean monthly values, vertical bars are the range of values, and horizontal bars are 1 standard error above and below the mean. From Hall, C. A. S. 1972. "Migration and Metabolism in a Temperate Stream Ecosystem." *Ecology* 53:585–604.

in ecosystem productivity in headwaters during the early spring coincided with the highest rates of fish migration to upstream reaches. Using energy budgets for both individual fish and the ecosystem as a whole, Hall calculated that the energetic costs of swimming upstream were rewarded by a roughly twenty-five-fold energetic benefit from feeding in a habitat with a higher amount of basal energy resources per unit volume. This energetic evidence that New Hope Creek fish were foraging optimally provided a quantitative explanation for upstream dispersal by stream biota.

What is perhaps most remarkable about this paper is how far ahead of its time it truly was on several fronts. It would be nearly three decades before another published study would document the seasonality of ecosystem energetics in a stream (see **Uehlinger 2000**), and it is only over the past few decades that we've seen attempts to directly investigate the efficiency with which basal energy resources are transferred into stream secondary production (Hall et al. 2003; Cross et al. 2013; Junker et al. 2020; Baruch et al. 2023).

Fisher, S. G., and G. E. Likens. 1973. "Energy Flow in Bear Brook, New Hampshire: An Integrative Approach to Stream Ecosystem Metabolism." *Ecological Monographs* 43:421–39.

In their 1973 *Ecological Monographs* paper, Stuart Fisher and Gene Likens reported the first comprehensive organic matter budget for a stream ecosystem. To do this, they applied the ecosystem mass-balance approach—only recently developed at the Hubbard Brook Experimental Forest (HBEF) in New Hampshire, USA (Bormann and Likens 1967; see chapter 10)—to Bear Brook, a stream draining one of the focal catchments within HBEF. Following this approach (*sensu* Bormann and Likens 1967; Likens and Bormann 1974), Fisher and Likens measured the meteorological, geological, and biological vectors by which energy entered the stream and the losses of that energy through respiration and physical export. They made no direct measurements of metabolic activity from diel gas concentrations and did not measure biological transport of energy by migrating animals in Bear Brook (as in **Hall 1972**). Instead, they made regular and comprehensive measurements of POM and DOM inputs, standing stocks, and exports.

Fisher and Likens discovered that Bear Brook was an ecosystem almost entirely supported by allochthonous inputs, or inputs of fixed carbon delivered from the surrounding catchment as leaves, twigs, and DOM. In this small headwater stream, animal biomass was extremely low, and yet nearly half of the carbon inputs to the study segment were transformed to dissolved organic carbon (DOC) or respired as $CO_2$. The stream transformed 34% of these energy inputs to heat via respiration and exported 66%. These values provided the basis for suggestions made later in the RCC that headwaters inefficiently leaked organic matter to downstream ecosystems. Fisher and Likens used the example of Bear Brook to question the popular assumption that ecosystems tend to approach a steady state with a GPP:ER (i.e., P:R) ratio of 1 as they develop (Odum 1969), and suggested that ecosystems with high organic matter inputs will have increased ecosystem efficiency as they develop, leading to lower GPP:ER ratios over time.

Decades of research and dozens of stream organic matter budgets later (many reviewed in Webster and Meyer 1997), we now know that most streams are indeed heterotrophic, and we can also say that Bear Brook in the late 1960s was among the most heterotrophic stream ecosystems ever studied. As influential papers are likely to do, this paper created a bias in the field that may have led to an overemphasis on the role of allochthony (see discussion in **Minshall 1978**) and an underappreciation of the role that animals can play in river energetics. Recent work has directed more attention to the critical role of instream autotrophy, even in highly heterotrophic ecosystems, in providing the kind of high-quality carbon inputs that are essential to sustain freshwater food webs (e.g., Thorp and Delong 1994; Marcarelli et al. 2011; Hotchkiss et al. 2015). In a somewhat amusing opposing trajectory, lentic system ecologists have recently undergone a similar awakening to the importance of allochthonous energy in supporting lake food webs that were too long assumed to be completely reliant on internal photosynthesis (Pace et al. 2007; Marcarelli et al. 2011; Tanentzap et al. 2017). More recent work has expanded the concept of allochthony to the ways in which organic matter inputs to streams are controlled by animals (Subalusky and Post 2019; see chapter 9).

Minshall, G. W. 1978. "Autotrophy in Stream Ecosystems." *BioScience* 28:767–71.

Minshall's 1978 paper pushed back against the growing assumption that terrestrial inputs largely support the metabolism and food webs of stream ecosystems (**Fisher and Likens 1973**; Cummins 1974). In that paper, Minshall pointed out the simple fact that not all streams are in deciduous forest landscapes; rather, some lie in terrestrial biomes that may have high light availability, and thus instream primary production dominates their organic matter budgets. As an example, Minshall highlighted Deep Creek, Idaho, USA—an open-canopy cool-desert stream—in which algal production far exceeded terrestrial leaf-litter inputs. Similarly, a colleague of Minshall's, C. E. Cushing, found almost pure autotrophy in a high-desert stream, Rattlesnake Creek, Washington, USA (Cushing and Wolf 1982, 1984). And we cannot forget that Odum's influential metabolism work occurred in a classically autotrophic stream (**Odum 1956**, 1957). Since Minshall's study, we have found that autotrophic streams and rivers are typical in deserts (Busch and Fisher 1981), in tailwaters below dams (Davis et al. 2012; Genzoli and Hall 2016), and in the South American Pampas (Acuña et al. 2011). Such ecosystems have high light availability and limited terrestrial inputs, while large dams typically sever downstream transport of organic matter.

Minshall's second key point was that just because a stream is "heterotrophic" does not mean that algae are somehow unimportant to food webs: in fact it is just the opposite. He posited, and many studies since have shown, that algae can support a substantial fraction of animal production despite constituting a small fraction of biomass and energy inputs into the stream at large. In particular, he identified the fallacy of assuming, just because an ecosystem has GPP:ER < 1, that terrestrial inputs fuel the food web (Rosenfeld and Mackay 1987; Meyer 1989). The outsized contribution of algae to food webs formed the basis of the riverine productivity model (Thorp and Delong 1994), and the few studies that have normalized the trophic basis of production using isotopes or gut contents analysis have shown that algae do contribute a larger fraction of production to food webs than might be assumed based on GPP:ER ratios (Huryn and Wallace 1985; Hall et al. 2001; McCutchan and Lewis 2002). Minshall's paper underscores the point that no matter how much carbon derives from the terrestrial landscape, instream photosynthesis may contribute greatly to a stream's carbon budget and to animal production.

Meyer, J. L., and R. T. Edwards. 1990. "Ecosystem Metabolism and Turnover of Organic Carbon along a Blackwater River Continuum." *Ecology* 71:668–77.

Meyer and Edwards used time series of metabolism from several sites along a blackwater river to test a core prediction of the RCC (**Vannote et al. 1980** [chapter 3]) and unify it with spiraling theory (**Newbold et al. 1981; Newbold, Elwood, O'Neill et al. 1983** [chapter 8]). The RCC laid out clear predictions of how GPP (and thus GPP:ER [or P:R]) would increase in mid-order rivers. Tests of these predictions focused on chamber estimates of metabolism in river continua throughout the United States (e.g., Minshall et al. 1983). Meyer and Edwards quantified how metabolism changed with time and season along a gradient of sites in the Ogeechee River, located in the Georgia coastal plain, USA. Their paper contains many firsts: it pioneered the use of reach-scale metabolism estimates (versus chambers) at several sites seasonally, allowing annual estimates of metabolism. Furthermore, the Ogeechee River was a true river continuum with almost no hydrologic or geomorphic alteration along its length (it is too flat to dam), and it had an intact riparian zone and floodplain. It was also a blackwater river with high concentrations of dissolved organic carbon (DOC). Meyer and Edwards found that this river was highly heterotrophic along its length, no doubt due to high loading of DOC from the floodplain. They also found that GPP and ER increased downstream, following a gradient of decreasing canopy coverage, as one would predict from the RCC.

Meyer and Edwards combined their organic carbon measurements with respiration data to examine organic carbon spiraling (Newbold et al. 1982) throughout the year at two sites. The spiraling length is the average distance traveled by organic carbon before it is respired to $CO_2$. Due to variation in discharge, organic carbon concentration, and respiration rates, they found huge variation in spiraling length, from a few hundred meters to over a thousand kilometers. One can compare this river to a "Slinky" (the spiral metal toy), with relatively high retention of organic carbon (and thus a shorter spiral) at low flows and low retention (and a longer spiral) at high flows, which act as a conduit for downstream transport. The authors' work presaged current efforts showing how flow controls where DOC is processed in catchments (Raymond et al. 2016; Manning et al. 2022) and the recent interest in linking riverine metabolism with carbon cycling (Plont et al. 2022; Battin et al. 2023). Another fun aspect of this paper is that Meyer and Edwards scaled metabolism to the entire drainage. Using simple hydraulic geometry relationships combined with a map, they showed that small streams contributed little to basin metabolism. This question of scaling river metabolism has generated intense current interest (Koenig et al. 2019), and using much more sophisticated computational methods, Segatto et al. (2021) also found that little of the network-level production came from the smallest streams within the Ybbs River network in Austria.

Finally, Meyer and Edwards pointed out the role of floodplains in river ecosystem functioning, reporting that even the highest-order stream in a continuum had high terrestrial organic matter loading if its floodplain was intact. This loading derived from lateral DOM inputs and episodic POM inputs during overbank floods (Meyer et al. 1997). They also demonstrated that lateral inputs from flood pulses (*sensu* **Junk et al. 1989** [chapter 9]) can be high even in large streams and rivers, a point not clearly made in the original formulation of the RCC.

Uehlinger, U. 2000. "Resistance and Resilience of Ecosystem Metabolism in a Flood-Prone River System." *Freshwater Biology* 45:319–32.

Urs Uehlinger was the first to take advantage of emerging sensor technologies to measure continuous variation in DO, and to use those data to estimate daily metabolic rates, in two hydrologically flashy subalpine rivers located in Switzerland throughout several years. While he had developed the techniques during short-duration deployments reported in an earlier paper (Uehlinger and Naegeli 1998), it is his 2000 *Freshwater Biology* paper that provided the first empirically measured stream metabolic regime. In addition to documenting seasonal variability (as was done in **Hall 1972**), Uehlinger's high-resolution dataset allowed him to focus on the extreme day-to-day variation in metabolic rates and to show that this variation was tied to bed-moving hydrologic disturbances. He used the recovery of rates of GPP and ER following storms to measure the resistance and resilience of stream autotrophs and heterotrophs to hydrologic disturbance. Through this work, he found that ER was far more resistant to bed-moving spates than was GPP, while GPP was more resilient, recovering more quickly following storms.

This paper brought stream ecosystem energetics into alignment with ongoing progress in the terrestrial ecosystem science community in developing eddy covariance techniques for making daily estimates of GPP and ER (Baldocchi 2020). The data revealed, once again, how distinct stream ecosystems are from most other ecosystems—this time in their extreme temporal variation in metabolic rates over short time scales. For streams, changes in metabolic rates over a storm-recovery time scale can exceed the variation seen across an entire year due to seasonal variation in light and temperature.

Uehlinger's demonstration of the primacy of hydrologic disturbance in constraining metabolic activity has implications for understanding how hydrologic alteration, via intentional water management or due to climate change, could have energetic consequences (Stanley et al. 2010; Finlay 2011). Ultimately, Uehlinger benefited greatly from the technical and financial investments of the Swiss National Hydrological and Geological Survey, which began monitoring DO concentrations alongside continuous discharge in the late 1990s. It would be more than a decade before other governmental agencies around the world implemented similar monitoring programs, which have enabled subsequent large-scale estimates of continuous stream metabolism (Appling, Read et al. 2018; Savoy et al. 2019; Bernhardt et al. 2022).

Fellows, C. S., H. M. Valett, and C. N. Dahm. 2001. "Whole-Stream Metabolism in Two Montane Streams: Contribution of the Hyporheic Zone." *Limnology and Oceanography* 46:523–31.

A major expansion of stream ecology theory was the recognition that a stream is not just the water visibly flowing in a channel, but also includes the extensive connection and exchange with near-stream groundwater—that is, the hyporheic zone (see chapter 2). Hyporheic zones have slow water flow and variable redox conditions, both of which facilitate biogeochemical transformations (**Triska et al. 1989b** [chapter 8]). Theory implies that the hyporheic zone's contribution to whole-stream function will be highest when flow from the channel to the hyporheic zone is high and when hyporheic biogeochemical transformations are fast (Findlay 1995). The problem is that it is difficult to quantify these hydrologic and biogeochemical fluxes. Fellows, Valett, and Dahm provided a clever solution to this problem.

Fellows and colleagues estimated the hyporheic contribution to reach-scale respiration in two montane streams in New Mexico, USA, that varied in the size of their hyporheic zones. They measured reach-scale respiration by using the **Odum (1956)** method, which integrates metabolic processes along surface and hyporheic flow paths. They then used chambers to estimate metabolism in the surface sediments and solved for the hyporheic metabolism by difference. The streams had roughly equal surface respiration, but one had hyporheic respiration that was ten times higher than its surface respiration. Their method followed that of Naegeli and Uehlinger (1997), but added solute tracer modeling that could hydrologically quantify the size of the hyporheic zone based on the short-term residence time of water, and indeed, they found that the site with the high hyporheic contribution had the longest water residence time in the hyporheic zone. The authors also calculated the physical size of each stream's hyporheic zone from the respiration measurements and found these much larger than the cross-sectional areas based on solute modeling.

Fellows and colleagues demonstrated how what we measure as respiration in a stream may not necessarily derive from surface sediments. Low-$O_2$ groundwater often re-emerges from the hyporheic zone, and it is difficult to determine whether the high apparent hyporheic contribution to respiration derives from high groundwater flow or high biogeochemical activity. This point can be tested directly via use of biogeochemically active tracers that can measure the flow of water in and out of the hyporheic zone, coupled with rates of reactivity based on tracer transformation (Haggerty et al. 2009). The authors showed that it is possible to quantify relative metabolic rates in stream transient storage zones. We also know that biogeochemical activity can be high in hyporheic zones: complete acetate removal in these same streams occurred along short flow paths (Baker et al. 1999). Fellows and colleagues' work provides a clear reminder that the boundaries of a stream extend far beyond the stream itself (**Fisher et al. 1998** [chapter 10]; Poole et al. 2008), and we suggest that much remains to be learned about hyporheic zone energetics.

## LOOKING FORWARD

The early conceptualization and measurement of ecosystem energetics began in flowing waters (**Odum 1956**), but streams did not remain at the forefront of this field. That is likely because many stream ecosystems possess unique features that make them exceptions to the assumptions and conceptual models that have influenced terrestrial ecosystem ecology. Living biomass in stream ecosystems is regularly removed by storms, violating steady-state assumptions (**Uehlinger 2000**). While in situ photosynthesis is a key resource for stream animals (**Minshall 1978**), most stream energy budgets are heavily subsidized by terrestrially derived organic matter (**Fisher and Likens 1973**), and higher subsidies are often associated with strong light limitation due to canopy shading (**Hall 1972**) or water column staining (**Meyer and Edwards 1990**). Streams have highly porous boundaries, not only with respect to interactions with their riparian zones and floodplains but also with respect to the local and regional groundwater with which they are hydrologically connected (**Fellows et al. 2001**). High disturbance frequency, heterotrophy, and connectivity are traits common to many of the best-studied stream ecosystems (see chapter 10), most of which are small streams that drain steep terrain. As research has expanded to include well-lit streams and larger rivers, we have discovered a much wider range of metabolic rates and regimes (Hall et al. 2015; Savoy et al. 2019; Bernhardt et al. 2022), and we are beginning to understand the spatiotemporal heterogeneity of production and organic matter retention within a wider range of river networks (Koenig et al. 2019; Segatto et al. 2021) using approaches pioneered by Naiman (1983) and **Meyer and Edwards (1990)**.

This expansion in the scope of our understanding of river metabolism has been enabled by a series of technical innovations. Early measures of metabolism in **Odum (1956)** and **Hall (1972)** relied on field sampling followed by labor-intensive Winkler titrations to measure dissolved oxygen. This work made estimates of metabolism incredibly time-consuming and impractical additions to many field campaigns. These logistical constraints have considerably delayed progress in incorporating ecosystem energetic measurements into stream food-web or biogeochemical studies, though papers that have made these links have been highly influential. For example, we know that stream metabolic rates can drive nitrogen uptake, transformation, and retention (Hall and Tank 2003; Roberts and Mulholland 2007; Mulholland et al. 2008). By comparison, we have fewer data that measure the extent to which ecosystem metabolism drives variation in secondary production, but hypothesize that we will see similarly strong links (Rüegg et al. 2020).

Since the 1990s, increasingly robust, field-deployable sensors have been able to measure DO nearly continuously throughout years of installation with only routine maintenance. Significant innovations in statistics, computing, and data science have facilitated the calculation of metabolism over long time scales across many sites. Now, we can run the **Odum (1956)** model for thousands of days of oxygen data from hundreds of rivers (Appling, Read et al. 2018). Yet few of the locations where oxygen is monitored are places where stream ecologists are measuring the identity, density, and mass of animals, except for a few intensively studied streams (e.g., Huryn and Benstead 2019; Hansen et al. 2023). Unfortunately, the simultaneous studies of ecosystem energetics and animal population dynamics (**Hall 1972**; **Fisher and Likens 1973**; Cuffney et al. 1990) that were central to the early development of stream ecosystem research have not continued. Making these links is a priority for our field going forward. When and how do stream animals become energy limited? How do animals navigate the spatiotemporal heterogeneity in energy availability in river networks? How are the answers to both of these questions changing as a result of river pollution, water management, and climate change?

Looking ahead, we see several opportunities and priorities for significantly expanding our understanding of stream energetics. The first among them is improvements and new capabilities in our metabolism models. Despite advances in inference and computation, we are still essentially using the **Odum (1956)** model unchanged, and much more sophisticated integration with hydrologic, geomorphic, and meteorological datasets and models to better account for physical processes should allow for greater accuracy and prediction of biological processes. The second area of high potential is the opportunity to link empirical measurements of metabolism with high-resolution satellite imagery for visible stream channels. With appropriate validation datasets, satellite estimates of stream color may soon provide reasonable estimates of productivity and organic matter fluxes that could be scaled to visible rivers across the globe (Ross et al. 2019; Gardner et al. 2021).

# 8

# NUTRIENT DYNAMICS

Daniel von Schiller, Eugènia Martí,
Marc Peipoch, and H. Maurice Valett

Nutrient dynamics in stream ecosystems addresses the inputs (e.g., diffuse and point sources), instream processing (e.g., assimilatory uptake, microbial immobilization, sorption), and losses (e.g., downstream transport, denitrification) of materials critical to the growth, maintenance, and reproduction of lotic biota. In this chapter, we provide a summary of the advancement of knowledge in stream nutrient dynamics and highlight foundational papers that have contributed exceptionally to this field. The selected papers provided key concepts and frameworks that have guided the field, with particular attention to the role of instream processes.

Early studies of stream nutrient dynamics focused on describing how nutrient concentrations in surface water changed with catchment modifications and nutrient inputs (Hynes 1960). Published works at that time tended to view streams as mere recipients of nutrients derived from surrounding terrestrial landscapes while generally ignoring instream processes. Nonetheless, in his foundational book, Hynes (1970) recognized the potential importance of instream biological processes, stating that "a lifeless river would have a very different chemical regime." More than fifty years of work since his book's publication has shown that streams are not simply pipes with no influence on the nutrients in their water, but instead are metabolically and biogeochemically active ecosystems in their own right.

An array of approaches contributed to early understanding of the influences of instream nutrient processes. Nutrient additions to streams provided evidence of nutrient limitation and uptake by stream organisms (e.g., McColl 1974; Stockner and Shortreed 1978). Longitudinal surveys similarly showed that instream processes could be responsible for downstream decreases in nutrient concentrations (e.g., Edwards 1974; Hill 1979). Phosphorus (P) radioisotope ($^{32}P$) releases allowed in situ measurement of P uptake and mineralization by stream biofilms (e.g., **Ball and Hooper 1963** [chapter 1]; Gregory 1978). Finally, budget (i.e., mass balance) studies, especially when combined with direct process measurements, revealed the significance of instream processes by comparing the forms and abundances of nutrient inputs and outputs (e.g., **Hall 1972** [chapter 7]; **Meyer and Likens 1979**; **Grimm 1987**).

The same combination of budget data and instream process measurements led **Webster and Patten (1979)** to coin the term "spiraling," initiating a concept that integrates nutrient uptake and downstream transport. In the early 1980s, researchers developed methods, metrics, and models that operationalized and expanded the spiraling approach (**Newbold et al. 1981**; **Newbold, Elwood, O'Neill et al. 1983**), inspiring an explosion of studies addressing and quantifying instream nutrient processes (e.g., Mulholland et al. 1985; **Triska et al. 1989a, b**). In the late 1980s, an effort to unify conceptual models and metrics for studies of instream nutrient processes produced a methodology paper (Stream Solute Workshop 1990) that foreshadowed the influence of this approach in subsequent years. Identifying the significance of nutrient transformations in the hyporheic zone was a key contemporary development (e.g., **Triska et al. 1989b**; Valett et al. 1990; Holmes et al. 1994), and several studies similarly showed that the riparian zone can be an important sink for nutrients in groundwater before they can enter streams (e.g., Lowrance et al. 1984; Peterjohn and Correll 1984). As techniques and approaches advanced, many studies in the 1990s (e.g., Nielsen et al. 1990; Pinay et al. 1993; Grimm and Petrone 1997) and early 2000s (Pinay et al. 2000; Kemp and Dodds 2001; Hall et al. 2002) added focus on the specific microbial transformations that underpin stream nutrient cycles.

The ability to release and measure nitrogen stable isotopes ($^{15}N$) in the field (Peterson et al. 1997), and to do so for dissolved N forms, was among the most significant developments in the investigation of N dynamics in streams. This approach was the basis of a large cross-site study known as the Lotic Intersite Nitrogen Experiment (LINX), which showed that instream uptake could control N concentrations in streams (**Peterson et al. 2001**). This approach was adapted to measure reach-scale nitrate uptake and denitrification in streams in a second large-scale study (LINX II). In that study, Mulholland et al. (2008) showed that the relative biogeochemical importance of small versus large streams is a function of nitrate loading: the importance of larger streams increased with greater loading rates as the capacity for uptake in smaller streams became saturated and a greater fraction of the nitrate load was exported downstream.

## HIGHLIGHTED PAPERS

Meyer, J. L., and G. E. Likens. 1979. "Transport and Transformation of Phosphorus in a Forest Stream Ecosystem." *Ecology* 60:1255–69.

Meyer and Likens's 1979 study, part of Meyer's dissertation, was the first to provide a comprehensive, whole-reach perspective on the dynamics of P in a headwater stream. Phosphorus is an essential element that, in the 1960s and 1970s, was shown to play a key ecological role in freshwater ecosystems because of its potential to limit rates of primary production (Schindler 1977; Stockner and Shortreed 1978). In this context, several studies during that era used $^{32}P$ as a radiotracer to assess the contributions of different instream biotic components to the assimilation of P from the water column (e.g., **Ball and Hooper 1963** [chapter 1]; Gregory 1978; Mulholland et al. 1985).

Meyer and Likens broadened the observational scope of P dynamics and were pioneers in quantifying reach-scale retention and transformation of natural inputs of P. Their study adopted the mass-balance approach used to assess elemental dynamics in catchment-scale studies (i.e., the small-catchment approach of Likens et al. 1977). Instream P retention and transformation was quantified as the difference between P input and export, or as the ratio of their fluxes, for different forms (dissolved, fine particulate, and coarse particulate) in a headwater stream reach. Intensive sampling over the course of a year yielded an annual P budget of unprecedented detail. Interpretation of the budget was broadened via the use of an existing thirteen-year dataset to examine how the instream P processing capacity varied over time within and among years.

Meyer and Likens's results highlighted the pulsed nature of P input and export and how this dictated the stream's capacity to retain P. Changes in the proportions of the different P forms between input and export also indicated the stream capacity to transform P, even when net P retention was negligible. Their study also highlighted the influence of hydrology on temporal variation of instream P retention—specifically, the pulsed transport associated with storms and the distinction between rising and falling limbs of spate hydrographs. Annual export:input ratios varied from <1 (P sink) to >1 (P source) and increased with annual streamflow. Despite these revelations at the reach scale, the methods employed did not allow examination of specific processes involved in P cycling. Moreover, P retention metrics derived from the budget approach were dependent on the length of the selected reach, a limitation addressed in later studies highlighted below.

The ideas generated in Meyer and Likens's work inspired substantial research over the following decades assessing nutrient dynamics, biogeochemical processing capacity, and their driving factors in stream ecosystems. Two novel developments that emerged from this study were (1) the identification of throughput or processing modes of stream biogeochemical reactivity and definition of the conditions that determine these modes, and (2) the idea that instream capacity to process bioreactive elements is a function of both hydrology and biological activity, which together determine nutrient demand. Meyer and Likens proposed water residence time, as well as the relationship between the stream's bioreactive surface and water volume, as factors controlling the proportion of P inputs that enter reactive biogeochemical pathways. They recognized that the influence of these features could vary with the bioavailability of different P forms and would hold for other bioreactive elements.

Webster, J. R., and B. C. Patten. 1979. "Effects of Watershed Perturbation on Stream Potassium and Calcium Dynamics." *Ecological Monographs* 49:51–72.

Behind a rather geochemical title resides one of the most influential studies in the field of stream nutrient dynamics. Webster and Patten's work introduced a novel concept, nutrient spiraling, to assess the dynamics of bioreactive elements in streams and rivers, ecosystems dominated by the unidirectional flow of water. The spiraling concept recognizes that the cycling of bioreactive elements occurs along a length of stream, a distance that is a product of instream processing capacity and hydrologic transport. Both the conceptual basis of the spiraling concept and its empirical mechanics have contributed significantly to the current view of streams as biogeochemical reactors, and the concept has increased our understanding of the role of streams in regulating nutrient inputs and outputs from catchments. In fact, the relevance of this concept has gone beyond the specific field of stream nutrient dynamics, as it is included in current textbooks on freshwater ecology and even general ecology. The concept conveys (1) that stream ecosystems are processors of nutrients, not just pipes, and (2) that spiraling length is an important trait of stream ecosystems that reflects both the characteristics of the ecosystem and how it influences nutrient export.

In addition to this conceptual contribution, Webster and Patten's 1979 paper highlighted the strong link between terrestrial and lotic ecosystems by examining the effects of catchment perturbations on streams (see chapter 9). They illustrated these effects by focusing on the dynamics of two essential elements, potassium (K) and calcium (Ca), in streams draining catchments with experimentally modified vegetation versus catchments kept unmodified as controls. Elemental dynamics were assessed by focusing on the content of K and Ca in riparian leaf litter, which itself acts as a major energy source for stream benthic communities (Fisher and Likens 1972). Differences in terrestrial vegetation among catchments altered the efficiency of stream biogeochemical processing; however, the study emphasized that streams can be highly resilient, recovering rapidly from terrestrial disturbance, and thereby reflecting the rapid turnover of instream components compared with rates of change in terrestrial vegetation.

Webster and Patten's work contributed significantly to the advancement of methodological approaches to studying stream nutrient dynamics from an ecosystem perspective. They developed field and laboratory methods, including radiotracers ($^{134}Cs$ and $^{85}Sr$), to assess dynamics of bioreactive elements in streams, focusing on different compartments (i.e., standing stocks) and processes (i.e., elemental fluxes among compartments). They also proposed a suite of metrics that facilitated stream functional characterization. Their work combined organismal and systems approaches by cleverly combining annual budgets at the ecosystem level (including input and output fluxes and material standing stocks) with rates of elemental trophic transfer to characterize each study stream and identify the effects of different catchment perturbations.

Newbold, J. D., J. W. Elwood, R. V. O'Neill, and W. Van Winkle. 1981. "Measuring Nutrient Spiralling in Streams." *Canadian Journal of Fisheries and Aquatic Sciences* 38:860–63.

Newbold, J. D., J. W. Elwood, R. V. O'Neill, and A. L. Sheldon. 1983. "Phosphorus Dynamics in a Woodland Stream Ecosystem: A Study of Nutrient Spiralling." *Ecology* 64:1249–65.

Two cornerstone papers by Newbold and colleagues advanced "spiraling" from a concept to an operational paradigm that directed studies of stream nutrient dynamics for the next forty years. The key contribution of their work was the development of a quantitative model to measure the average distance traveled by a nutrient atom to complete one full cycle, which they named the "spiraling length" (figure 8.1). Their goal was to include both dissolved and particulate phases in a mechanistic approach to quantifying spiraling of nutrients in streams, building on the ideas of **Webster and Patten (1979)** and pioneering $^{32}P$ additions in streams (**Ball and Hooper 1963** [chapter 1]; Elwood and Nelson 1972). But it was the ability of Newbold and colleagues to reduce complex processes to simple rules that allowed them to conceive such a fundamental measure of nutrient dynamics in streams.

These two papers are complementary analyses of the same $^{32}P$ release in the West Fork of Walker Branch that Newbold organized during his time at Oak Ridge, Tennessee, USA. In the 1981 paper, Newbold and his colleagues first partitioned the spiraling length among the distances traveled by P atoms dissolved in the water (i.e., uptake length), in suspended particles, and in consumer compartments (i.e., turnover length). Their results showed that transport in the consumer compartment was negligible compared with that in dissolved or particulate form, which led to the contemporary view of spiraling length as the sum of uptake and turnover length. Over the years, studies of spiraling have typically characterized only the uptake length, a bias that can likely be traced back to the paper's operational estimation of turnover length as a relatively small and constant proportion (roughly 13%) of the spiraling length. However, only two years later, the 1983 paper provided a more

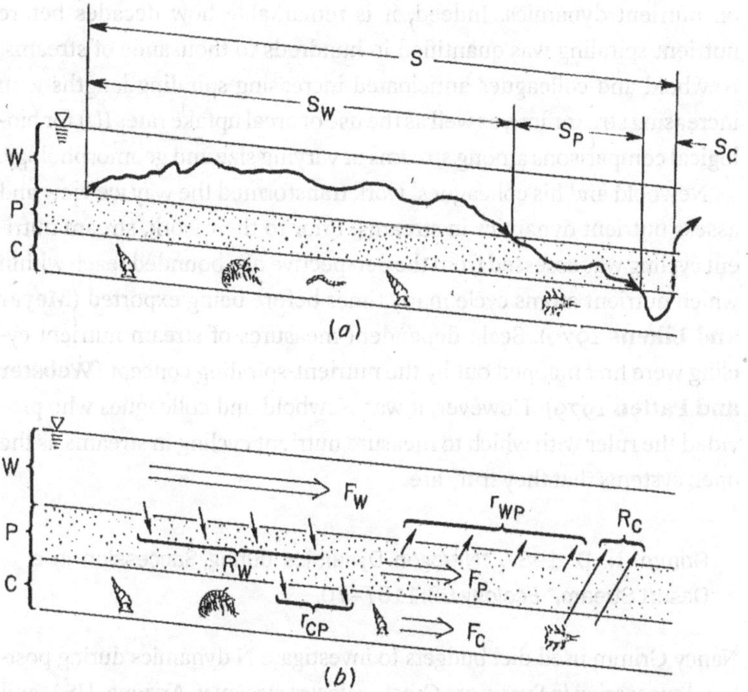


Fig. 8.1 Nutrient spiraling in a stream ecosystem. (*a*) The three components of spiraling length, *S*. The nutrient atom travels an average longitudinal distance, $S_W$, dissolved in the water compartment, *W*, plus an average distance, $S_P$, in the particulate compartment, *P*, plus an average distance, $S_C$, in the invertebrate consumer compartment, *C*, before returning to the water. (*b*) Downstream nutrient fluxes, *F* (g/s), and exchange fluxes, *R* (g/s/m), of nutrients between compartments. The total release from the particulate compartment, $R_P$, consists of the sum of the release to the water, $r_{WP}$, and the uptake by consumers, $r_{CP}$. From Newbold, J. D., J. W. Elwood, R. V. O'Neill, and W. Van Winkle. 1981. "Measuring Nutrient Spiralling in Streams." *Canadian Journal of Fisheries and Aquatic Sciences* 38:860–63.

detailed model that characterized large variation in nutrient turnover across twelve different biological compartments. By doing so, Newbold and colleagues provided the original quantification of nutrient spiraling. Amazingly, this study stands as the most complete assessment of spiraling to date, including spiraling estimates in the stream water and in the food web. With their model, Newbold and colleagues showed that only select stream biota are actively involved in nutrient cycling. The paper and its authors were also prescient in terms of future perspectives on nutrient dynamics. Indeed, it is remarkable how decades before nutrient spiraling was quantified in hundreds to thousands of streams, Newbold and colleagues anticipated increasing spiraling lengths with increasing streamflow, as well as the use of areal uptake rates (*U*) for biological comparisons among streams of varying size and geomorphology.

Newbold and his colleagues' work transformed the way we view and assess nutrient dynamics in streams. Prior to their work, stream nutrient cycling was assessed from the perspective of a bounded reach within which nutrient atoms cycle many times before being exported (**Meyer and Likens 1979**). Scale-dependent measures of stream nutrient cycling were first mapped out by the nutrient-spiraling concept (**Webster and Patten 1979**). However, it was Newbold and colleagues who provided the ruler with which to measure nutrient cycling in streams as the open systems that they truly are.

Grimm, N. B. 1987. "Nitrogen Dynamics during Succession in a Desert Stream." *Ecology* 68:1157–70.

Nancy Grimm used diel budgets to investigate N dynamics during postflood succession in Sycamore Creek, a desert stream in Arizona, USA, and a landmark site for lotic research (see chapter 10). Grimm's 1987 study addressed fundamental theory related to ecosystem energetics, succession, and nutrient retention. In addition, it was the first budget for desert streams, whose characteristics (extreme hydrologic variability, high autotrophic productivity, and strong N limitation) make them ideal models to study how N dynamics change with disturbance and succession.

With inputs to the reach exceeding outputs on most study dates, Sycamore Creek exhibited net N retention. Marked diel changes in N concentrations, together with the mass-balance results, showed that retention, mostly of inorganic N, was controlled by autotrophic assimilation. These results reinforced the idea of streams as ecosystems in which a strong coupling between nutrient uptake and net ecosystem production generates a high capacity for nutrient retention. In this case, most N retention was in benthic algae (~90%), but a significant amount was also retained in macroinvertebrates (~10%). In fact, in a follow-up study, Grimm (1988) further demonstrated the significant contribution of stream macroinvertebrates to ecosystem N cycling. Overall, Grimm's results fueled interest in a better understanding of the importance of instream uptake for catchment nutrient budgets, a topic of great interest in biogeochemistry during the following decades (e.g., Mulholland 1992; Bernhardt et al. 2003; Brookshire et al. 2009). Furthermore, in the discussion section of her 1987 paper, Grimm pointed to other processes besides assimilatory uptake that could contribute to N cycling (e.g., $N_2$ fixation, denitrification, and hyporheic processes), which paved the way for subsequent studies on the relevance of these processes for N dynamics in lotic ecosystems (e.g., Grimm and Petrone 1997; Boulton et al. 1998; Mulholland et al. 2008).

Grimm's study was also one of the first to use streams to investigate theory relating disturbance to nutrient cycling. Specifically, her study explicitly tested Vitousek and Reiners's (1975) nutrient retention hypothesis developed for forest ecosystems. This hypothesis suggests that nutrient retention reflects net production: retention is initially low following forest disturbance, increases from early to middle successional stages in association with plant establishment and growth, then declines in the later stages of succession as biomass accretion slows. Grimm's results showed that patterns of algal growth and N retention following flash floods in desert streams are well described by this model, emphasizing autochthonous links between nutrient dynamics and ecosystem energetics. Later studies revealed that, while this model holds for autotrophic desert streams, it might not hold for allochthonous-based temperate forest streams (Valett et al. 2002). In addition to their general contribution to ecosystem science, the insights in this paper have implications for understanding the response of stream ecosystem functioning to restoration (Palmer and Ruhi 2019).

Triska, F. J., V. C. Kennedy, R. J. Avanzino, G. W. Zellweger, and K. E. Bencala. 1989a. "Retention and Transport of Nutrients in a Third-Order Stream: Channel Processes." *Ecology* 70:1877–92.

Triska, F. J., V. C. Kennedy, R. J. Avanzino, G. W. Zellweger, and K. E. Bencala. 1989b. "Retention and Transport of Nutrients in a Third-Order Stream in Northwestern California: Hyporheic Processes." *Ecology* 70:1893–905.

In 1989, less than a decade after the publication of the foundational papers operationalizing nutrient spiraling, a group of five researchers from the United States Geological Survey generated two sequential papers that provided additional conceptual and operational understanding of nutrient retention, which would occupy the forefront of lotic ecology in the coming decades.

Solute release experiments (Stream Solute Workshop 1990), which began in the 1980s but remained abundant and influential through the next twenty-five years, represented a transformative method for studying nutrient cycling in whole stream reaches and their integrated subsystems. However, most such studies were short-term, lasting only a few hours. Thus, Triska and colleagues' study, which involved the co-injection of a hydrologic tracer (chloride) and a biologically dynamic solute (nitrate) in a headwater stream, was among the longest (twenty-one days) and most spatially extensive (327 m) solute releases conducted to that point. As the authors wrote (Triska et al. 1989a), "reliable ecosystem-level measurements are rare due to enormous problems of scale, particularly in physical factors." While this experiment is notable due to its spatial and temporal scale, it is the elegant integration of physical factors with biological counterparts that places this pair of studies within the foundational realm.

The question that drove the five researchers, "Is nitrate concentration in stream water a product of localized cycling within a reach or a measurement of passive transport from the terrestrial environment?" (Triska et al. 1989a), anticipated a decade of research addressing the role of streams in catchment function. While the first paper (Triska et al. 1989a) described the temporal patterns of biotic uptake and transient storage over the course of the experiment, the second (Triska et al. 1989b) proposed an expanded view of lotic ecosystem boundaries in the hope of identifying subsurface storage zones, determining the spatial and temporal character of surface-subsurface exchange, and characterizing the tendency of these subsurface environments to process reactive N species.

These two papers stand as robust treatises that together represent a veritable guidebook for solute co-injections. Their schematic presentations (e.g., figure 8.2) and mathematical representations provide the template for both theoretical and empirical assessment of mechanisms governing transport and retention in streams. Their schematic representation of the hyporheic zone as a biogeochemical reactor organized by exchange flows with the channel stands as a robust and ongoing hypothesis that continues to foster research. The lasting and far-reaching character of these papers is a projection of the interdisciplinary nature of the consortium of authors, which included ecologists, engineers, and hydrologists. Together, their ideas and approaches anticipated decades of research and hundreds of papers addressing groundwater–surface water interaction (Jones and Mulholland 2000), transient storage zone modeling (Runkel and Chapra 1993; Kelleher et al. 2019), and the emergence of ecohydrology as a field (Nuttle 2002).

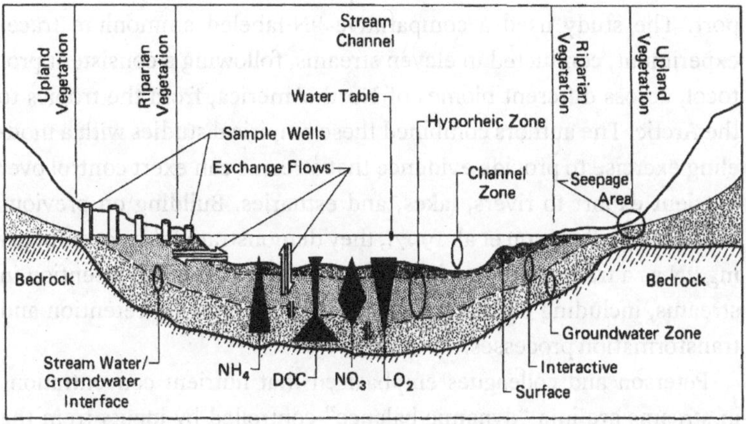


Fig. 8.2 Conceptual model of the groundwater–surface water linkage at Little Lost Man Creek. Waters are divided into three zones: a channel zone containing surface water, a hyporheic zone, and a groundwater zone. The hyporheic zone has been divided into a surface hyporheos with virtually identical chemistry to channel waters, and containing >98% advected channel water, and an interactive hyporheos characterized by physical-chemical gradients (e.g., $NH_4$, $O_2$, DOC, $NO_3$, $O_2$, temperature). The interactive zone consists of <98% but >10% advected channel water. The overall solute concentration is theoretically represented by the shape of the concentration profile. Transport of solutes across the groundwater–stream water interface (the stream boundary) is a function of the hydrologic head of the adjacent landscape or response to a concentration gradient, whereas transport from the channel zone to the hyporheic zone is dominated by advective processes. From Triska, F. J., V. C. Kennedy, R. J. Avanzino, G. W. Zellweger, and K. E. Bencala. 1989b. "Retention and Transport of Nutrients in a Third-Order Stream in Northwestern California: Hyporheic Processes." *Ecology* 70:1893–905.

Peterson, B. J., W. M. Wollheim, P. J. Mulholland, J. R. Webster, J. L. Meyer, J. L. Tank, E. Martí, W. B. Bowden, H. M. Valett, A. E. Hershey, W. H. McDowell, W. K. Dodds, S. K. Hamilton, S. Gregory, and D. D. Morrall. 2001. "Control of Nitrogen Export from Watersheds by Headwater Streams." *Science* 292:86–90.

While streams had historically been sampled at the bases of catchments to reflect changing terrestrial influences on elemental uptake and transport (e.g., Vitousek and Reiners 1975), Peterson and colleagues executed a groundbreaking effort employing isotopic tracers and a spiraling approach across multiple lotic systems to definitively address the potential for instream processes to modify material export. The study used a comparative $^{15}N$-labeled ammonium tracer experiment, conducted in eleven streams, following a consistent protocol, across different biomes of North America, from the tropics to the Arctic. The authors combined these empirical studies with a modeling exercise to provide evidence that streams can exert control over nutrient export to rivers, lakes, and estuaries. Building on previous studies (e.g., Peterson et al. 1997), they demonstrated the power of using $^{15}N$ as a tracer to unravel nutrient transformation and retention in streams, including permanent removal and temporary retention and transformation processes.

Peterson and colleagues emphasized that nutrient concentrations in streams are in a "dynamic balance" controlled by inputs from the catchment and instream uptake and release processes. Moreover, they recognized that the role of streams in controlling nutrient export might be limited over annual cycles because streams are not net accumulators of nutrients. Their study suggested, however, that permanent removal of nitrogen via N-gas emissions due to coupled nitrification and denitrification might be relevant. Peterson and colleagues also proposed that small streams, characterized by shallow depths and high surface-to-volume ratios, are more efficient than larger streams in taking up nutrients (Alexander et al. 2000). This fact has been contested by more recent studies showing high rates of nutrient uptake in large rivers (Hall et al. 2013; Wollheim et al. 2022).

The use of $^{15}N$ release techniques, and the capacity to measure the abundance of isotopes in dissolved form accurately, allowed Peterson and colleagues to show that ammonium was taken up through several processes, including sorption, assimilation, and nitrification, with implications for its ultimate fate. Discerning the fate of the added tracer allowed for assessment of the differing roles of stream compartments (e.g., epilithon, benthic organic matter, macrophytes) in N uptake, measurement of the flow of N through stream food webs, and some of the first in situ measurements of eventual remineralization (e.g., Mulholland et al. 2000; Tank et al. 2000). Importantly, the work showed that a significant portion of N uptake was remineralized within a short period, contributing to the aforementioned "dynamic balance" of stream nutrient concentrations. Subsequent studies in the frame of the LINX II project using additions of $^{15}$N-labeled nitrate demonstrated the importance of denitrification (Mulholland et al. 2008) and associated $N_2O$ emissions (Beaulieu et al. 2011) to the removal of N from streams.

Finally, Peterson and colleagues' work represents one of the first examples of a large-scale collaborative stream experiment using standardized protocols across biomes, and it trained a generation of graduate students who have gone on to make major contributions to the study of stream biogeochemistry. This type of study allows results to be generalized beyond a single site, and this paper inspired several international stream research initiatives conducted years later (e.g., Boyero et al. 2011; LINX Collaborators 2014; Datry et al. 2018; Tiegs et al. 2019).

## LOOKING FORWARD

Nutrient dynamics is a relatively new field in stream ecology, with foundations set in the last part of the twentieth century; many knowledge gaps remain that represent opportunities for conceptual growth. Understanding how nutrients enter streams, how they are processed within these ecosystems, and how streams can control export of nutrients to downstream ecosystems is a vibrant field of research. Along these lines, more recent developments in the study of stream nutrient dynamics have included a closer look into the influence of physical processes on instream N removal (Grant et al. 2018), the development of new methods for measuring instream nutrient dynamics (Covino et al. 2010; Heffernan and Cohen 2010), and the extension of these approaches to larger rivers (Tank et al. 2008).

As in many other scientific fields, methodological and technical developments have been crucial for advancing knowledge in stream nutrient dynamics. Just as the development of whole-stream nutrient-enrichment and isotope-tracer methods was key to research through the early twenty-first century, the use of high-frequency nutrient sensors is setting the path for current advances in the field (Heffernan and Cohen 2010; Rode et al. 2016). As these sensors become more affordable, accurate, and able to measure new analytes (e.g., stable isotopes), they will allow a more precise characterization of nutrient dynamics in stream ecosystems. The related development of better models at the scale of stream networks and whole catchments that incorporate stream nutrient transport and cycling (Helton et al. 2011; Wollheim et al. 2022) will allow a better understanding of stream nutrient dynamics and their role on a large scale.

The study of stream nutrient dynamics cannot rely solely on methodological and technical developments, however; it should also build on conceptual advances (Fisher et al. 2004). One of the areas in which combining new methodological capabilities and key conceptual advances can produce the most valuable output is the concomitant assessment of stream nutrient dynamics from alternative observational frames (fixed location, Lagrangian, and synoptic samplings; Ensign et al. 2017). Similarly, the coupling of nutrient and carbon dynamics within a stoichiometric perspective is an important conceptual framework for advancement of biogeochemical understanding (Hall and Tank 2003; Schade et al. 2011; Williamson et al. 2016). Finally, recent conceptual developments are aimed at incorporating nutrient dynamics in intermittent streams characterized by both wet and dry phases (Gómez-Gener et al. 2021). The field of stream nutrient dynamics will continue to benefit from a good balance among conceptual, methodological, and technical developments.

It is not easy to foresee the future directions of a research field. However, it is likely that researchers in the field of stream nutrient dynamics will continue to address unresolved questions recognized as relevant to the field of ecology as a whole. For example, still under debate is the extent to which streams can exert control over the export of nutrients to rivers, lakes, and estuaries. This issue remains relevant because of the potential importance of this ecosystem service: the self-cleansing character of streams has been purported for some time (e.g., **Patrick 1949** [chapter 1]). Related to this issue is the question of the effectiveness of large rivers, as compared with streams, in removing nutrients and the limited information on turnover lengths compared with that on uptake lengths (see **Newbold et al. 1981; Newbold, Elwood, O'Neill et al. 1983;** Tank et al. 2008). Understanding turnover requires accurate measures of mineralization, and the concept's application to larger rivers becomes more relevant as more of their organic matter stocks are found to be actively transported. Addressing these areas requires improved approaches to measuring nutrient uptake in large rivers, as well as accurate differentiation between gross and net nutrient uptake. Such advances will produce more and better empirical data from streams and especially from large rivers. These data can then be used to feed process-based nutrient budget models that incorporate the biogeochemical role of stream networks, from small streams to large rivers, in whole catchments (e.g., Helton et al. 2011).

Restoration is a common practice aimed at recovering healthy streams and rivers. A better understanding of the drivers of nutrient dynamics in well-preserved streams, and how they change with stream perturbation, will contribute to better stream restoration guidelines that consider not only biodiversity, but critical ecosystem functions as well. In addition, most stream studies to date have focused on the dynamics of N and P. Future studies may turn to other less studied macro- and micronutrients whose dynamics, especially in the context of instream processing, are far less understood.

# 9

# RECIPROCAL STREAM-LAND INTERACTIONS

Colden V. Baxter, Amanda Subalusky, and Hiromi Uno

Interactions between land and water are a critical essence of stream ecology, as is evidenced by the treatment of such interactions in every other chapter of this volume. Beginning with the field's origins (see chapter 1), empirical studies and conceptual models that helped define the discipline have demonstrated that stream ecosystems are inextricably connected to their valleys and riparian zones and that, owing to these connections, changes on land (e.g., human land uses, pollutants) can lead to changes in streams. Work by **Hynes** (1970, **1975** [chapter 2]) and others (e.g., Likens and Bormann 1974) demonstrated early on that changes in the valley are accompanied by changes in the stream; these demonstrations included the paired-catchment experiment led by the latter authors that showed how deforestation could dramatically alter stream water chemistry (see chapter 10). This background seeded the themes of subsequent conceptualizations of stream habitats (see chapter 2), communities, and ecosystem processes (see chapter 3). Early empirical studies revealed the role of terrestrially derived organic matter sources in sustaining aquatic macroinvertebrates (see chapter 1), echoing a long-standing appreciation that fishes often consumed food delivered from the land (Baxter et al. 2005). The idea of the land (especially forested landscapes) "feeding" streams was a rich one that grew to shape investigations and understanding of food webs (see chapter 5) and organic matter processing (see chapter 6), as well as ecosystem energetics (see chapter 7) and nutrient dynamics (see chapter 8). While early studies focused on the importance of leaves shed seasonally by streamside vegetation to organisms like macroinvertebrates, investigations of other pathways and complex consequences of land-water interactions have helped set the stage for a more holistic, ecosystem perspective on streams, their adjacent riparian zones, and their entire catchments (**Gregory et al. 1991**). Researchers found that fluxes of dissolved organic molecules can constitute a large portion of catchment inputs to streams (see chapter 7). Concurrently, studies of streams and floodplain rivers in forested catchments raised awareness of the importance of large wood as habitat for organisms (Keller and Swanson 1979) and wood's mediation of various ecosystem processes (Bilby 1981). This observation underscored the consequences of historical practices of channelizing and removing wood from streams, rivers, and their floodplains (**Sedell and Froggatt 1984**; Wohl 2014). Studies also revealed that relations between the vegetation and the hydrogeomorphology of river corridors are reciprocal (**Gurnell et al. 2001**). Moreover, terrestrially derived organic matter was not only important to sustaining stream macroinvertebrates, but had extensive indirect effects on communities across trophic levels (Wallace et al. 1997; Nakano et al. 1999).

Inputs to streams from land were conceptualized as most important in headwater contexts for many years, notably via the river continuum concept's (RCC) iconic conceptual figure (**Vannote et al. 1980** [chapter 3]). In subsequent decades, investigations of larger river-floodplain and tributary-mainstem interactions have contributed to a more expansive appreciation of the mosaic-like pattern of land-water connectivity. Key work on the former was contributed by researchers working on large undammed rivers in tropical regions (**Junk et al. 1989**) and relatively intact river floodplains remaining in temperate zones (e.g., **Tockner et al. 2000**; Stanford et al. 2005). Understanding of land-water connectivity was also extended when the ecological importance of exchanges of materials and organisms between ground and surface waters was revealed by European investigators (see Dole-Olivier et al. 1994), then developed and assimilated into the field's conceptual paradigms in the 1980s and 1990s (Stanford and Ward 1993).

For decades, studies concentrated on one-way fluxes from land to water, but much recent work has turned attention to fluxes from streams to land, especially via movements of organisms such as emerging adult aquatic insects (Jackson and Fisher 1986; Sabo and Power 2002) or via translocation of organisms and organic matter between water and land mediated by animals like amphibians (Regester et al. 2006). Important early papers illustrated how these processes may be engineered by animals such as beavers (**Naiman et al. 1986**) or facilitated by various predators and scavengers on aquatic organisms like salmon (**Willson et al. 1998**). In turn, such work led to a conceptualization of streams and land as reciprocally linked (**Nakano and Murakami 2001**) in a "tangled web" of relationships, explicit investigation of feedback interactions, and growing awareness that streams and the land are interwoven in their complexity, dynamism, and vulnerability to agents of global change (Baxter et al. 2005). In this chapter, we highlight a combination of conceptual and empirical papers from a range of geographic contexts, all chosen to illustrate the foundations and development of our understanding of reciprocal stream-land interactions, and for their influence on contemporary and future research.

## HIGHLIGHTED PAPERS

Sedell, J. R., and J. L. Froggatt. 1984. "Importance of Streamside Forests to Large Rivers: The Isolation of the Willamette River, Oregon, USA, from Its Floodplain by Snagging and Streamside Forest Removal." *Internationale Vereinigung für theoretische und angewandte Limnologie: Verhandlungen* 22:1828–34.

While the origins of stream ecology (see chapter 1) and foundational concepts like the RCC (**Vannote et al. 1980** [chapter 3]) emphasized the importance of riparian forests for small streams, Sedell and Froggat's elegant 1984 study drew attention to the value of forests located along large rivers and on their floodplains. By conducting an analysis of historical records and constructing and comparing a time series of maps, they revealed that historical removals of large woody debris ("snagging") and cutting of floodplain forests over decades had strikingly reduced the complexity of the floodplain and simplified channel configurations of the Willamette River, Oregon, North America (figure 9.1). This work made a strong case that streamside forests interact with landforms and play an important role in maintaining river-floodplain connectivity. Furthermore, the authors warned that most mid-sized to large rivers in the temperate zone around the world have experienced a legacy of alteration, and that because much of their loss of habitat complexity and connectivity occurred long ago, their original "baseline" may have been forgotten. This phenomenon was later emphasized by other investigators of large river-floodplain ecosystems (**Junk et al. 1989**) and has become increasingly apparent in the years since (Wohl 2014).

This paper was produced as part of a series of efforts by an interdisciplinary team of ecologists, geomorphologists, and natural resource managers to re-evaluate the value of streamside and floodplain forests in the Pacific Northwest of the United States, a region with massive native forests and socially important fish species in decline. The series of works drew researchers' attention to the role of landforms in rivers and amplified appreciation for streamside forests and wood in rivers around the world. Sedell and Froggat's study played an early part in highlighting the role of floodplain plants as engineers of riverine habitat

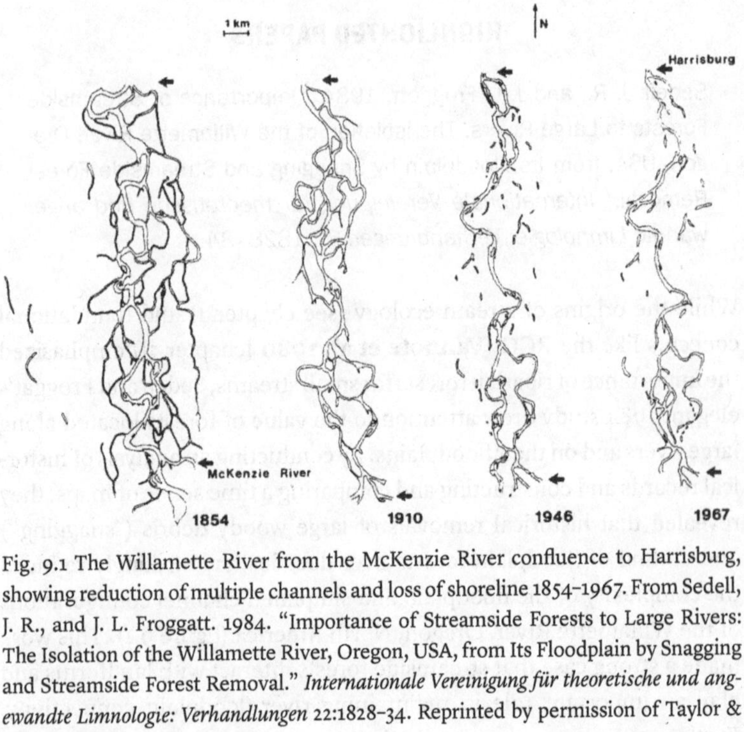


Fig. 9.1 The Willamette River from the McKenzie River confluence to Harrisburg, showing reduction of multiple channels and loss of shoreline 1854–1967. From Sedell, J. R., and J. L. Froggatt. 1984. "Importance of Streamside Forests to Large Rivers: The Isolation of the Willamette River, Oregon, USA, from Its Floodplain by Snagging and Streamside Forest Removal." *Internationale Vereinigung für theoretische und angewandte Limnologie: Verhandlungen* 22:1828–34. Reprinted by permission of Taylor & Francis Ltd.

(**Gurnell et al. 2001**), and drove home the implications of their loss. The study also foreshadowed decades of work that has revealed the reciprocal processes by which forests are maintained by, but also create, heterogeneous, dynamic river-floodplain landscapes. Contemporary researchers describe these processes with such phrases as the "shifting habitat mosaic" (Stanford et al. 2005) and the "floodplain large-wood cycle hypothesis" (Collins et al. 2012).

Naiman, R. J., J. M. Melillo, and J. E. Hobbie. 1986. "Ecosystem Alteration of Boreal Forest Streams by Beaver (*Castor canadensis*)." *Ecology* 67:1254–69.

The significant role that animals can play in connecting land and water, and in driving aquatic ecosystem processes, was highlighted in Naiman, Melillo and Hobbie's 1986 paper on the role of beavers in boreal streams. Their analysis demonstrated how the density of beaver dams in natural ecosystems could drive stream geomorphological processes and biogeochemical cycling. They synthesized a range of measurements, including allochthonous inputs from the catchment and atmosphere, instream organic matter cycling rates, and outputs via carbon export and insect emergence, across a range of stream orders with and without beaver activities. They then constructed carbon budgets for an average stream riffle versus a beaver pond. They showed that beaver activities directly and indirectly increased sediment retention and small and large woody debris inputs in streams, decreased decomposition rates, and increased carbon export downstream, highlighting beaver ponds as sites of "substantial organic matter storage and processing."

This paper connected beaver activities to the growing interest in the role of large woody debris in stream and river-floodplain ecosystems (**Sedell and Froggatt 1984**; **Gregory et al. 1991**) and examined how animal movement could shape its spatial and temporal dynamics, thus increasing stream spatial heterogeneity. Naiman and colleagues also quantified the influence of stream microhabitat (e.g., riffles and pools) on biogeochemical cycling and illustrated how the input of woody debris could modify biogeomorphological processes (**Gurnell et al. 2001**). The characterization of this heterogeneity contributed some much-needed breadth to the conceptualization of headwater streams as treated in the RCC (Burchsted et al. 2010). This perspective also influenced understanding of the role of habitat heterogeneity in shaping ecosystem processes, leading, for instance, to investigations of "hot spots and hot moments" of biogeochemical cycling (McClain et al. 2003).

The paper contributed to a broader understanding of the potential role of animal movements in influencing stream processes. Naiman and colleagues specifically recognized beavers as a keystone species, which connected their study to ongoing work on this phenomenon in streams and rivers, as well as the emerging concept of ecosystem engineers (Wright et al. 2002). It also helped lay the groundwork for examining the influence of beaver ponds and meadows on species diversity, community assembly, and community stability (e.g., McDowell and Naiman 1986), as well as the ecological legacies of the loss of beavers (Wohl 2021).

Junk, W. J., P. B. Bayley, and R. E. Sparks. 1989. "The Flood Pulse Concept in River-Floodplain Systems." *Canadian Journal of Fisheries and Aquatic Sciences* Special Publication 106:110–27.

The "flood pulse concept" proposed by Junk, Bayley, and Sparks accelerated acceptance of land-water connectivity as important to the dynamics of lotic ecosystems and drew attention to its importance in river-floodplain settings. Their 1989 paper was a response to the RCC (**Vannote et al. 1980** [chapter 3]), whose tenets and predictions did not reflect the observations of ecologists studying large river-floodplain ecosystems. Junk and colleagues drew on a growing literature (including much on fish ecology; e.g., Goulding 1980) and their own results from studies of relatively unaltered rivers of the Neotropics and Southeast Asia as well as sites on the upper Mississippi River, contexts where the seasonal flood pulse had not been lost to human modifications of flow regime or floodplain connectivity. Based on various lines of evidence and their synthesis of the literature, they argued that, in the context of unaltered river segments, the influence of the flood pulse mattered far more than longitudinal positioning, such that a "flood pulse concept" was needed to complement the RCC. They described how predictable high flows inundate floodplains seasonally, creating a "moving littoral" or "aquatic-terrestrial transition zone." This seasonal flooding, they argued, facilitates lateral exchange and recycling, such that the bulk of riverine organism productivity is derived from this river-floodplain interaction rather than the main river channel, and traits of communities reflect adaptation to this lateral connectivity rather than to longitudinal gradients.

Beyond its proposal of a complementary concept to the RCC, this paper raised awareness that the natural land-water interface of large rivers is marked by shifting ecotones rather than rigid boundaries (a point reiterated by **Gregory et al.** [**1991**] and others). It also amplified appreciation for the significance of lateral movements in river-floodplain ecosystems by a wide range of aquatic and amphibious animals, echoing the observations of **Naiman et al.** (**1986**) regarding beavers and anticipating what has become an increasingly important area of study. Subsequently, segments of rivers with extensive, connected floodplains have come to be conceptualized as distinct "functional process zones" (Thorp et al. 2006) marked by dynamic lateral and vertical mosaics that are characteristic of river environments, rather than exceptions to a rule (e.g., Stanford et al. 2005). Moreover, given the disproportionate influence of studies on small temperate streams, the paper remains a bridge between the ecology of big rivers and small streams, between tropical and temperate contexts, and between fisheries science and river ecology. It also challenges stream ecology as a discipline, because understanding connected, complex river-floodplain ecosystems requires study of both lotic and lentic habitats—which together form ecosystems whose dynamism defies the classification historically separating limnological subdisciplines.

Gregory, S. V., F. J. Swanson, W. A. McKee, and K. W. Cummins. 1991. "An Ecosystem Perspective of Riparian Zones." *BioScience* 41:540–51.

A pivotal paper by Gregory, Swanson, McKee, and Cummins pushed scientists and natural resource managers to adopt more holistic perspectives on stream-land linkages. Drawing on studies from the Pacific Northwest and synthesis of the broader literature, Gregory and colleagues highlighted the importance of the land-water interface to lotic ecosystems and argued that understanding and management of riparian zones should reflect an appreciation for the dynamic processes that create and maintain their mosaic of habitats. They proposed "a conceptual model of riparian zones that integrates the physical processes that shape valley-floor landscapes, the succession of terrestrial plant communities on these geomorphic surfaces, the formation of habitat, and the production of nutritional resources for aquatic ecosystems." Like **Naiman et al. (1986)** and **Junk et al. (1989)**, they focused on the "ecotonal" nature of this interface, emphasizing that the interface should not be treated as a rigidly delineated or static boundary. Their paper reflected growing appreciation of natural disturbances not as disasters, but as keys to ecological integrity and as connections between riparian ecology and natural flow regimes. The authors presented data depicting the patchiness of geomorphic settings in riparian zones at hierarchically nested scales (**Frissell et al. 1986** [chapter 2]), showing how this pattern hosted biodiversity and created heterogeneity in terms of quantity and quality of organic matter inputs to streams (later termed "resource subsidies"; see **Nakano and Murakami 2001**).

The paper made spatial complexity itself both a focus of study and a trait to manage, rather than treating it as noise around a theoretical ideal, foreshadowing the integration of landscape ecology and "metacommunity/ecosystem" frameworks into stream ecology. The authors emphasized the role of riparian zones in the spatial ecology of organisms (e.g., as corridors for plant dispersal and animal movement) and the reciprocity of water-land fluxes (e.g., sediment, nutrients, and organic matter are deposited laterally on floodplains as well as being transported into and down streams). They also articulated feedbacks such as the influence of landform on riparian vegetation states and that of vegetation on geomorphic patterns and processes (**Sedell and Froggatt 1984**; **Gurnell et al. 2001**). Though few stream ecologists would now dispute the need for "an ecosystem perspective of riparian zones," many of the same challenges to which these authors were responding persist, and the basic messages of the paper remain highly relevant.

Willson, M. F., S. M. Gende, and B. H. Marston. 1998. "Fishes and the Forest." *BioScience* 48:455–62.

While much research on land-water connections focused on inputs from terrestrial ecosystems to streams, Willson, Gende, and Marston's

1998 paper on the influence of Pacific salmon on terrestrial ecosystems raised fundamental questions about the importance of resources moving in the other direction. A growing body of research during this time was demonstrating the important role of Pacific salmon in streams through transport of marine-derived resources and their incorporation into freshwater food webs (Bilby et al. 1996; Wipfli et al. 1998). Several studies had also documented the importance of salmon carcasses to terrestrial consumers (Willson and Halupka 1995; Ben-David et al. 1998). Willson and colleagues drew together this body of work to argue that anadromous fishes, including not only Pacific salmon but also the lesser-known char, smelt, herring, and sand lance, provide a "rich, seasonal food resource that directly affects the biology of both aquatic and terrestrial consumers and indirectly affects the entire food web that knits the water and land together."

The authors provided an overview of the abundance and distribution of anadromous fishes in inland systems and their importance to terrestrial consumers, arguing that anadromous fishes are keystone species for vertebrate consumers (Willson and Halupka 1995). They also explored the fertilizer effects of salmon carcasses that are deposited on land, whole or in excreta, and their incorporation into soil, riparian vegetation (e.g., Helfield and Naiman 2001), and higher trophic levels, and they estimated the potential magnitude of these fluxes in the phosphorus budgets of various regional systems. Finally, the authors explored the implications of these linkages for fisheries, forestry, and wildlife management, highlighting the need for integrative management approaches that consider impacts across ecosystem boundaries.

This framing, illustrated by the tangled aquatic and terrestrial food webs depicted in figure 1 of their paper, built on ideas posed by Goulding (1980) about the interactions between Amazonian fishes and flooded forests and provided a prescient view of the future study of aquatic and terrestrial exchanges. Indeed, a substantial body of research on salmon during and after this period increasingly recognized their effects on both aquatic and terrestrial ecosystems and the reciprocal interactions between them (Cederholm et al. 1999; Gende et al. 2002), and thus stimulated more study of marine-freshwater and other downstream-to-upstream connections (Pringle 1997).

Tockner, K., F. Malard, and J. V. Ward. 2000. "An Extension of the Flood Pulse Concept." *Hydrological Processes* 14:2861–83.

Gurnell, A. M., G. E. Petts, D. M. Hannah, B. P. Smith, P. J. Edwards, J. Kollmann, J. V. Ward, and K. Tockner. 2001. "Riparian Vegetation and Island Formation along the Gravel-Bed Fiume Tagliamento, Italy." *Earth Surface Processes and Landforms* 26:31–62.

While the flood pulse concept proposed by **Junk et al.** (**1989**) became a major concept in the study of floodplain rivers, it was principally founded on evidence and observations from tropical lowland rivers. Tockner, Malard, and Ward (2000) and Gurnell and colleagues (2001) pointed out that many temperate floodplain rivers have characteristics that differ from their tropical counterparts, and thus extended the flood pulse concept to these temperate ecosystems. While most European rivers have been heavily altered, European researchers, in an effort to understand the seminatural baseline of river ecosystems in the region, have conducted intensive, multidisciplinary research on the relatively intact Tagliamento River in Italy. In this context, Gurnell and colleagues made intensive field observations and described how living and dead riparian vegetation, sediment, and hydrologic regime interactively form spatially heterogeneous, multithreaded channels with many islands. A main point of the extension of the flood pulse concept by Tockner and colleagues was the focus on the roles of spatial heterogeneity supported by such river-riparian interactions.

The authors of these two linked papers described expansion-contraction cycles driven by pulses of flow well below flood stage as a physical characteristic of floodplains, and described how those cycles support habitat heterogeneity and functional processes. They also pointed out high spatial variability in water temperature as a major controller of floodplain ecology. Ecological functions of river-riparian spatial heterogeneity and its dynamism (often described as a "shifting habitat mosaic"; Stanford et al. 2005) have become a major research arena for river ecologists. Such spatial heterogeneity not only enables the coexistence of diverse species and life stages (e.g., Arscott et al. 2005), but also plays important roles in species interactions (e.g., Uno 2016). Moreover, the growing body of work that links the physical and hydrologic processes of river and river-floodplain interactions has informed flow management and river restoration projects around the world (Stanford et al. 1996; Nakamura et al. 2006).

> Nakano, S., and M. Murakami. 2001. "Reciprocal Subsidies: Dynamic Interdependence between Terrestrial and Aquatic Food Webs." *Proceedings of the National Academy of Sciences* 98:166–70.

Perhaps more than any other, Nakano and Murakami's 2001 study supported adoption of a truly reciprocal vision of stream-land interactions. The paper drew on rigorous observations over two years of the deciduous-forested Horonai Stream in northern Japan. The authors and their team measured fluxes of terrestrial invertebrates to the stream and of emerging adult aquatic insects to the forest, biomass of stream and forest invertebrates, and abundance of and feeding by fishes and riparian birds. Remarkably, they showed that nearly half of the energy budget of the fish assemblage was sustained by consumption of terrestrial invertebrates, and that over 25% of the energy need of the entire bird assemblage was fueled by emerging insects (figure 9.2). Moreover, emergence peaked in spring and provided a consistent, low-level flux to the forest throughout the winter, both periods when terrestrial invertebrate biomass was low, whereas terrestrial invertebrate input to the stream peaked in summer, when the biomass of benthic stream insects was low. This asynchronous phenology meant that the forest "fed" stream fishes in summer, and the stream "fed" all forest birds in spring as well as year-round resident birds in winter. This dynamic interdependence meant that the forest could sustain more birds, and the stream more fish, than could have been supported without these cross-habitat fluxes of prey.

The study was a breakthrough, but one tragically marked by the death of Shigeru Nakano months before its publication. Nakano and colleagues (1999) had already published an experimental study demonstrating how inputs of terrestrial invertebrate prey from the forest altered cascading trophic interactions in the stream, such that when these fluxes were cut off using a replicated greenhouse-style manipulation, stream fish reduced benthic invertebrates, causing an indirect increase in streambed periphyton. Subsequently, another experiment showed that the reduced benthos also caused reduced insect emergence back to the riparian forest, resulting in fewer riparian spiders (Baxter et al.

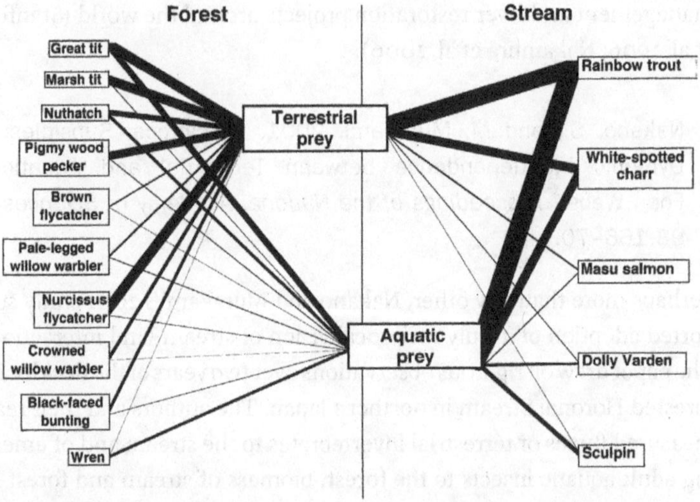


Fig. 9.2 Food-web linkage across a forest-stream interface representing predator subsidies by allochthonous, invertebrate prey flux. Relative contributions of terrestrial and aquatic prey to the annual total resource budget of each species were represented by line thickness. Annual resource budget of each species was represented as a value proportional to the total assemblage-based budgets, separately for the bird and fish assemblages. From Nakano, S., and M. Murakami. 2001. "Reciprocal Subsidies: Dynamic Interdependence between Terrestrial and Aquatic Food Webs." *Proceedings of the National Academy of Sciences* 98:166–70. Copyright (2001) National Academy of Sciences, U.S.A.

2004). These observations were complemented by findings from a third replicated greenhouse experiment, in which a single 1.2 km reach was covered and responses tracked over multiple years (reviewed by Fausch et al. 2010), and by a recent analysis (Marcarelli et al. 2020) that drew on data from Nakano and Murakami's 2001 paper to reveal that indirect effects of terrestrial invertebrate inputs on the stream food web changed with time scale. Thus, the landmark 2001 paper is accompanied by a rich "ecology of place" (see chapter 10), yielding compelling evidence for the importance of reciprocal stream-land interactions.

## LOOKING FORWARD

The study of land-water connections has been a foundational theme in stream ecology, and has contributed to coupled management of rivers and riparian forests. However, many important directions for future research on these and related topics remain.

First, there is a relative paucity of studies that address reciprocal exchange of resources between land and water within the same ecosystem. Indeed, despite a burst of "resource subsidy" research over the past twenty years, most investigations have focused on consequences of fluxes in one direction or the other, and there have been few investigations (cf., Collins et al. 2020) of the actual reciprocity of such linkages. This focus belies the truly intertwined relationship between streams and their catchments. Both abiotic (flooding, hyporheic flow) and biotic (aquatic insect emergence, terrestrial animal foraging) vectors can move resources from streams into the landscape. These aquatic subsidies can influence productivity in riparian ecosystems, which in turn can influence productivity in streams. There is a need for increased research on the spatial ecology of reciprocal stream-land linkages that extends beyond simple, lateral connections to include variations in stream network topology and floodplain mosaics. There is also an opportunity to apply perspectives gained from reciprocal stream-land interactions to other aquatic connections such as mainstem-tributary, stream-lake, and stream-ocean linkages. Such work may inform more holistic approaches to both land and water resource management that prioritize the conservation and restoration not only of terrestrial and aquatic systems, but also the connections between them.

Second, little research focuses on variability in the characteristics of reciprocal land-water inputs and how this variability mediates consequences for organisms and ecosystem processes (Subalusky and Post 2019). Major axes of such variability include the quantity and quality of resources (Marcarelli et al. 2011). Although many studies have quantified the magnitude of fluxes across ecosystem boundaries, resource quality may play an equally important role in determining effects in recipient ecosystems. For example, net carbon fluxes from terrestrial to aquatic systems are typically several orders of magnitude larger than net carbon fluxes from aquatic to terrestrial ecosystems (Gounand et al. 2018); however, aquatic subsidies often have high fatty acid concentrations, and thus can provide critical resources for reproducing songbirds and other riparian consumers (Twining et al. 2019). Also, despite their centrality to **Nakano and Murakami's (2001)** findings, there has been relatively little research on the importance of the temporal dynamics and complementarity of reciprocal linkages between streams and land, though it is apparent that subsidy timing can vary widely within and among ecosystems (Rundio and Lindley 2012; Uno 2016). Changes in the phenology of subsidies in terms of timing or duration may introduce mismatches between consumers and resources that can have significant direct and indirect effects on community structure and ecosystem processes, and such changes are likely to become more common and more pronounced under climate change (Larsen et al. 2016). Direct and indirect effects of these linkages may also lead to different responses across different time scales (Marcarelli et al. 2020), highlighting the need for longer-term studies that explicitly incorporate land-water exchanges.

Third, there is a need for study of reciprocal stream-land linkages across a broader array of environmental contexts. Though studies have now been conducted in settings ranging from small desert streams (Jackson and Fisher 1986) to river-floodplain ecosystems (Paetzold et al. 2006), and from Mediterranean climates (Sabo and Power 2002) to coastal rainforests of the Pacific Northwest of North America (Marczak and Richardson 2007), most work has continued to be focused on perennial, temperate stream ecosystems. Though research on intermittent streams has increased in the past decade (Datry et al. 2014), they remain understudied with respect to reciprocal stream-land linkages. In terms of geographic variation, streams in different parts of the world are shaped by different temperature and precipitation regimes likely to influence the form and timing of both aquatic and terrestrial fluxes (e.g., Chan et al. 2008). Increasing research on these topics in tropical, desert, and boreal ecosystems will broaden understanding of potential drivers and context dependencies in reciprocal land-water exchanges. This broader lens will also result in research on a greater taxonomic range of organisms involved in mediating water-land-water linkages. Beavers and salmon provide classic examples of animals that act as keystone species and ecosystem engineers in perennial, temperate freshwater systems; however, other regions of the world host different species that may play similar or distinct functional roles (Subalusky and Post 2019). Broadening the geography of stream ecology research will also broaden the diversity of researchers engaged in this work, a critically important goal for the future of the field.

# 10

# THE ECOLOGY OF PLACE

Barbara L. Peckarsky and Mary E. Power

Stream ecology has been enriched by an "ecology of place" approach (Billick and Price 2010): the prolonged study, often by several generations of researchers, of one site and its environs. Deep understanding of place enables researchers to build on fundamental knowledge of natural history, connections forged with other local knowledge holders, and methods developed by those who came before. Place-based research often attracts researchers from different disciplines with diverse ideas and perspectives. Similarly, field stations, even humble ones, facilitate more prolonged, focused, or serendipitous conversations among researchers who share interest in a site than is typically possible at offices or laboratories on busy campuses. Guided by observations accumulated over time at one place, researchers can design and implement surveys and experiments at times, locations, and scales suitable for resolving the mechanisms producing ecological patterns and driving their changes over time. New discoveries may expand on or challenge previous findings to generate more comprehensive understanding. By extending the "history" in natural history, years of experience at one location enable researchers to recognize gradual change (e.g., in seasonal phenology, hydrologic regimes, biota), or to detect abrupt ecosystem shifts and then test ideas about why they are happening. Prolonged studies of ecosystems that document change and identify its causes are invaluable, particularly as global change accelerates.

Place-based research can expand to more general understanding when subsequent findings force investigators to resolve "surprises" (Lindenmayer et al. 2010; Dodds et al. 2012): unexpected or counterintuitive findings that arise either because ecological "rules" that applied during previous years have been bent or broken by environmental or biotic change, or because repeated observations reveal the influence of processes overlooked in smaller datasets. The need to resolve those inconsistencies is both the challenge and the beauty of working in one place. While careful scientists always address this challenge, there is a well-known bias for confirmatory results across all scientific publications (Joober et al. 2012). Yet science moves forward when new, unexpected outcomes question cherished hypotheses. We hope to reassure students and younger stream ecologists that it is not just OK, but beneficial, to observe something unexpected.

In this chapter, we describe five long-term, place-based research programs that have contributed to the foundations of stream ecology.* While insights in stream ecology also arise from short-term studies at multiple places, knowledge of one specific place provides a fundamental understanding of river and stream ecology that can be generalized well beyond the site at which it was developed. This chapter includes not only scientific discoveries, but also narratives of how one discovery led to the next to build the current level of understanding of how each system works. We asked investigators how their thinking changed or evolved over the years, whether they encountered surprises, and if so, how such surprises were resolved by continuing work. Learning about the experiences and capturing the growing, synthetic insights of

---

* Space constraints precluded our covering the fundamental contributions to stream ecology made by many long-term place-based research programs. A number of those sites are in the US National Science Foundation Long Term Ecological Research Network [e.g., H. J. Andrews Experimental Forest (Oregon), Konza Prairie Biological Station (Kansas), Kuparuk River/Toolik Field Station (Alaska), and the Luquillo Experimental Forest (Puerto Rico)]. Other sites include La Selva Research Station in Costa Rica, several South Island streams in New Zealand (see chapter 5), Oak Ridge National Laboratory (Tennessee, USA; see chapter 8), the Ogeechee and Satilla Rivers (Georgia, USA; see chapter 7), the Nyack River floodplain (Montana, USA), Tai Po Kau Nature Reserve in Hong Kong, Broadstone Stream in the UK (see chapter 5), the Breitenbach in Germany, and the Tagliamento River in Italy (see chapter 9).

long-term place-based researchers enabled us to better understand how each place specifically, and ecology of place in general, has contributed to the field of stream ecology.

Selecting five places was not easy. We chose sites with contrasting hydrologic regimes, elevations, drainage areas, and vegetation types that were broadly distributed across the continental United States (figure 10.1). We included two sites where we have worked ourselves: the Eel River and its tributaries in the Northern California Coast Range, and the East River and its high-elevation tributary streams in western Colorado. In addition, we summarize long-term research in the forested headwater streams at the Hubbard Brook Experimental Forest in northern New Hampshire; in Sycamore Creek, a Sonoran Desert stream in south-central Arizona; and in forested streams in and around the Coweeta Hydrologic Laboratory in the southern Appalachians of western North Carolina. Studies at all of these sites shared several characteristics: (1) investigator-driven research began with immersion in natural history; (2) questions arising from observations were addressed by experiments that revealed underlying ecological mechanisms; and (3) investigators confronted surprises that they then sought to explain.

## HIGHLIGHTED PLACES

Hubbard Brook Experimental Forest: Bernhardt, E. S., G. E. Likens, R. O. Hall Jr., D. C. Buso, S. G. Fisher, T. M. Burton, J. L. Meyer, W. H. McDowell, M. S. Mayer, W. B. Bowden, S. E. G. Findlay, K. H. Macneale, R. S. Stelzer, and W. H. Lowe. 2005. "Can't See the Forest for the Stream? The Capacity of In-Stream Processing to Modify Terrestrial Nitrogen Exports." *BioScience* 55:219–30.

Gene Likens began his epic research journey at Hubbard Brook Experimental Forest (HBEF) in 1961, collaborating with Herb Bormann (Bormann and Likens 1967) to pioneer one of the most influential catchment-scale experiments ever conducted in forested headwater streams (Likens et al. 1970; Likens et al. 1978). A small (0.16 km$^2$) headwater catchment was deforested, and herbicide was applied to prevent the re-establishment of vegetation for three years. Subsequent sampling of the clear-cut catchment, along with a reference, uncut catchment, revealed export of nitrate from the former so massive that Likens,

Fig. 10.1 Locations of place-based studies described in this chapter. (1) Rocky Mountain Biological Laboratory, East River and tributaries: high elevation, snowmelt streams on western slope of Colorado Rocky Mountains; elevation 2,863–3,216 m, latitude 38.96, longitude –106.99, drainage basin area 69.9 $km^2$; protected field station on US Forest Service (USFS) land. (2) Hubbard Brook and tributaries (*top*): temperate deciduous forested headwater streams in northern New Hampshire; elevation 220–1,015 m, latitude 43.94, longitude –71.69, drainage basin area 30 $km^2$; protected field station on USFS land. (3) Angelo Coast Range Reserve, upper South Fork of the Eel River and tributaries (*bottom*): Mediterranean hydrology in the Northern California Coast Range; elevation 300–1,420 m, latitude 39.73, longitude –123.63, drainage basin area 330 $km^2$; protected by the University of California Natural Reserve System. (4) Coweeta Hydrologic Laboratory, Coweeta Creek and tributaries (*overleaf top*): forested headwater streams in the southern Appalachian Mountains of western North Carolina; elevation 675–1,592 m, latitude 35.06, longitude –83.43, drainage basin area 16.3 $km^2$; protected field station on USFS land. (5) Sycamore Creek (*overleaf bottom*): lowland Sonoran Desert stream in south-central Arizona; elevation 427 m, latitude 33.63, longitude –111.66, drainage basin area 505 $km^2$; no field station; USFS land. Photos courtesy of (1) A. R. McIntosh, (2) Emily Bernhardt, (3) Mary Power, (4) Phillip Bumpers, and (5) Nancy Grimm.

Fig. 10.1 (*continued*)

Fig. 10.1 (*continued*)

surprised, sent the samples to many different laboratories to confirm that vegetation loss could exert such profound effects on biogeochemical fluxes through streams (Robert Hall, personal communication). The concept that "the valley rules the stream," from the classic paper "The Stream and Its Valley" (**Hynes 1975** [chapter 2]), was appropriately attributed to the original work of Likens and his colleagues at HBEF. This experiment changed stream ecology "from typically reductionist and biologically focused to a more integrative and holistic discipline. No modern stream ecologist would draw a bright line at the water's edge" (David Strayer, personal communication).

Sustained research at HBEF illustrates how place-based research fosters collaborations among scientists in different disciplines. As **Hynes (1975)** stated, "Some of our most important recent discoveries have been of the existence of hydrologists, foresters and soil scientists." The Hubbard Brook Ecosystem Study, initiated in 1963, was an interdisciplinary effort including Likens (limnologist), Bormann (plant ecologist), Robert Pierce (soil scientist) and Noye Johnson (geologist). As a PhD student, Likens became smitten by the interdisciplinary ecosystem approach, which he emphasized with his students, because it integrates diverse perspectives needed to understand complex ecosystems. As he told one of those students, "The opportunity to learn from researchers in fields such as hydrology, geomorphology, forestry and soil science drew me to the field, and has kept me there" (Meyer 2013).

Years of observations and experiments at HBEF have produced many surprises. For example, streams and forests may respond differently to disturbance, even following opposite trajectories (Hall 2003). Consequently, long-term changes in terrestrial catchments may not predict long-term biogeochemical changes in receiving streams. Similarly, Kate Macneale discovered unexpected cross-catchment dispersal of adult stoneflies away from their natal streams through the forest to adjacent streams, revealing the potential for cross-basin gene flow among aquatic insects at HBEF (Macneale et al. 2005) and elsewhere (Hughes et al. 1999; Briers et al. 2004).

The 2005 study, which includes contributions from most of Likens's HBEF PhD students, illustrates how the resolution of counterintuitive observations can lead to new insights about ecosystems. Synthesis of forty years of observational and experimental studies at multiple temporal and spatial scales explained surprising regional declines in stream nitrate concentration and export that countered expectations based on traditional models of terrestrial ecosystem nutrient cycling (Vitousek and Reiners 1975). In their synthesis of previous research at HBEF, the investigators ruled out alternative explanations and highlighted the role of instream processing of N to explain the regional declines. This synthesis culminated in a new model describing long-term changes in instream nitrogen cycling as forests age (figure 5 in Bernhardt et al. 2005), which proposed mechanisms (e.g., increased algal uptake of nitrate and its longer-term storage in stream consumers) that have implications for in-channel controls of N concentration and transport through ecosystems. The accumulated research at HBEF shows the strength and reciprocity of links between "the stream and its valley," demonstrating that catchment models must include stream processes as important components.

The atmosphere at the HBEF field station (Pleasant View Farm) fostered continuity among generations of stream ecologists. Former students at HBEF developed a strong sense of community as an important aspect of place-based ecology. Generations of HBEF researchers built on earlier research to investigate the biogeochemical integration of streams with their catchments, from Likens to Fisher to Bilby to Meyer to McDowell to Sloane to Hedin to Hall to Bernhardt to Rosi to Marinos (see Table 13.1 in Holmes and Likens 2016). Bernhardt credits Likens with creating an environment where place-based research could have its greatest impact by building "intellectual connective tissue," fostering links between cohorts by encouraging students to read the work of their predecessors. Dissertations, theses, and all publications derived from research at HBEF were stored in the library at the ramshackle farmhouse, which initially served as housing and a rudimentary lab, providing an environment where students "constantly talked science, over meals or games of horseshoes or in the garden" (David Strayer, personal communication). Shared ideas and voluntary participation as extra field hands built a supportive atmosphere. While he worked on lakes, Strayer knew what the stream ecologists were doing, and he asserted that he (or anyone else) could have given the talks of other HBEF researchers in their cohort at scientific meetings.

Unresolved inconsistencies remain at HBEF. For example, although subsequent research at HBEF reported algal blooms during spring snowmelt (Bernhardt and Likens 2004), **Fisher and Likens (1973** [chapter 7]) reported negligible algae in Bear Brook following a winter with extensive anchor ice, concluding that forested headwater streams were heterotrophic, and thereby contributing to the long history of research on detritus and heterotrophy in forested streams (e.g., chapter 1, chapter 6, **Wallace et al. 2015**). In the dark headwater streams at HBEF, 95% of the overall primary consumer production is supported by heterotrophic energy sources (Hall et al. 2001). However, Mayer and Likens (1987) observed high densities of algae-grazing caddisflies in Bear Brook. What were they eating? Algae made up less than 2% of the organic matter in the stream channel, but represented 50% of their gut contents and an estimated 80% of the assimilated energy that fueled them. Possible explanations are that minuscule amounts of algae can support dense grazers due to high food quality and fast turnover rates (Vadeboncoeur and Power 2017), and that caddisflies may be much better at collecting algae than stream ecologists are (Jacques Finlay, personal communication).

Foundational studies at HBEF over many decades reflect Likens's appreciation of place and his approach to testing hypotheses at multiple scales and levels of observation. His place-based research enabled subsequent generations of students not only to solve local mysteries reported by their predecessors, but also to carry his legacy of insights and approaches from HBEF to other places, including our next site.

Sycamore Creek: Fisher, S. G., N. B. Grimm, E. Martí, R. M. Holmes, and J. B. Jones Jr. 1998. "Material Spiraling in Stream Corridors: A Telescoping Ecosystem Model." *Ecosystems* 1:19–34.

Stuart Fisher and Nancy Grimm discovered ecosystem science in New England as a student and "grand-student," respectively, of Gene Likens. In their foundational Bear Brook paper, **Fisher and Likens (1973** [chapter 7]) had adapted a mass-balance approach—pioneered for lakes by Lindeman (1942) and for catchments by Likens et al. (1970)—to a New Hampshire stream. This same concept, and an appreciation for place-based ecology, inspired four decades of research by Fisher, Grimm, and their students and colleagues in a very different ecosystem, Sycamore Creek, a lowland Sonoran Desert stream in south-central Arizona, USA.

Desert streams are among Earth's most dynamic ecosystems: Australians describe theirs as "dry or ten foot high." Winter rains and summer monsoons produce catastrophic floods that obliterate biota, then rapidly recede to dry up portions of the channel (Fisher et al. 1982), motivating fundamental questions best addressed by long-term place-based research: What controls post-flood recovery of biota? How does biotic succession proceed? How are the biogeochemical cycles underpinning ecological recovery re-established after profound and frequent disruption? What are the temporal and spatial scales of heterogeneities governing biogeochemical flow paths and "hot spots" and "hot moments" of biogeochemical transformations (McClain et al. 2003)? To address those questions, Grimm, Fisher, and their students have tracked sources, sinks, and transformations of fluxing solutes, sediments, and biota, not only across heterogeneous patches on the flowing streambed, but also in interacting sub-compartments below the streambed (the hyporheos) and "all the way to the divide" (Fisher, North American Benthological Society Award of Excellence lecture, 1998).

While early ecosystem ecologists promoted catchments as a natural scale for studying ecological responses to change, much of stream ecology has primarily focused on the wetted channel. Research at Sycamore Creek was notable in extending beyond these traditional system boundaries. Ryan Sponseller assumed he would be working in streams when he began his PhD research at Arizona State University. "Instead, I spent most of my dissertation working with plants and soils, trying to understand how the broader drainage network organized ecosystem patterns and processes across the landscape." Others in his cohort (Jill Welter, Jennifer Edmunds) also studied desert uplands as extensions of Sycamore Creek connected/disconnected by precipitation events. Students were encouraged to investigate the importance of the surprising hydrologic connections between the stream and the uplands revealed by early studies at Sycamore Creek, advancing long-term, place-based research and leading researchers to address questions that incorporated new spatial extents or new parts of the channel network.

Sycamore Creek research contributed a new conceptual framework to the foundations of stream ecology by encompassing complex fluxes and transformations through three "ecosystem compartments": the channel, the hyporheic zone, and an "air force reserve" of ovipositing aquatic insects (Gray and Fisher 1981) that recolonized flood-scoured streams. The fluxes of emerging insects, which sustain terrestrial animals (Jackson and Fisher 1986), motivated an explosion of interest in cross-ecosystem fluxes ("resource subsidies") connecting streams and upland catchments (**Nakano and Murakami 2001** [chapter 9]; Power et al. 2004; Baxter et al. 2005). Transport and connectivity across other "compartment" boundaries were also key to understanding stream ecosystem processes. Nutrient-rich hyporheic water enters the surface stream at upwelling sites, fueling algal recovery after floods (Valett et al. 1994). At downwelling sites, DOC in the water column is imported to hyporheic zones, supporting microbes within sediments (Jones et al. 1995a, b). Groundwater flow paths similarly connect the surface stream to lateral subsystems. For example, $NO_3$ accumulates from mineralization and nitrification along streamside gravel bars (Holmes et al. 1994), supplying reactive nitrogen to the stream at "in-welling zones," which in turn can shape the spatial arrangement of $N_2$-fixing and non-$N_2$-fixing algae along the reach (Henry and Fisher 2003). Finally, reciprocal hydrologic connections also extend to riparian vegetation, which can serve as an important nitrogen sink when these lateral groundwater connections are strong (Schade et al. 2005).

In their 1998 study, Fisher and his colleagues used decades of place-based research to wrestle such complex spatial dynamics into a comprehensible model. Extending the simpler nutrient-spiraling theory (**Newbold et al. 1981** [chapter 8]), their "telescoping ecosystem model" for material spiraling through stream corridors envisioned a fluvial ecosystem as three nested concentric cylinders: the flowing stream (from streambed to water surface); the hyporheic zone (water, solutes, matter, and life below the streambed), and the adjacent riparian zone. With changes in loading, loss, and transformation rates, those cylinders, like tubes in a telescope, extend or contract in a downstream direction and relative to each other. The length of each cylinder varies inversely with the loss rate of substances advected through it due to chemical transformation or net flux out of the cylinder. Factors such as time since disturbance, spatial redox gradients, or patchy biotic uptake or removal control material processing rates. The material processing length in the telescoping ecosystem model is analogous to the uptake length in spiraling theory: it is the average distance traveled by a specific molecule before it is chemically transformed or removed from advective transport. Calibrating and validating this conceptual model required mapping of spatial variation in rates across small (redox gradients) to large (geomorphic, climatic, geological) scales, as well as tracking the system state over time relative to resets by disturbance and successional recovery.

Fisher and his colleagues operationally defined the lengths and time scales of key process components of the telescoping ecosystem model (see their Table 1), allowing its broad application across humid, mesic, semi-arid, and desert streams for quantitative comparisons of ecosystem responses to landscape or climate changes.

**Stanley et al. (1997** [chapter 3]) suggest that, despite its challenges, place-based research on desert streams offers several advantages for advancing foundational understanding of stream ecosystem dynamics. Flashy desert hydrographs show ecosystem responses to hydrologic forcing over wide ranges of variation. In warm deserts, high productivity accelerates succession, so recovery from flood scour (Fisher et al. 1982) and drought (**Stanley et al. 1997**) can easily be examined within "dissertation" or "NSF grant" time scales. But above all, the long history of research and mentoring provided by senior scientists at Sycamore was key to meeting research challenges.

Former students of Fisher have emphasized the advice they received to "spend as much time as possible at Sycamore Creek without agenda: observe the system, learn the 'players,' and think about questions while being out there—rather than from behind the computer" (Ryan Sponseller, personal communication). Fisher's mentoring included legendary Wednesday evening lab meetings where students discussed papers, interacted with visiting speakers, and talked science, from general philosophy to methodological approaches. "Fisher thought deeply about these things, e.g., where do good ideas come from? (e.g., Fisher 1997). And he shared these musings with us, much to our long-term benefit" (Emily Stanley, personal communication). Sycamore Creek's legacy lives on even as its practitioners disperse and apply their four-dimensional thinking to Arctic, Mediterranean, coastal forest, and urban catchments around the world.

Angelo Coast Range Reserve: Power, M. E., M. S. Parker, and W. E. Dietrich. 2008. "Seasonal Reassembly of a River Food Web: Floods, Droughts, and Impacts of Fish." *Ecological Monographs* 78:263–82.

Reprising the theme that long-term place-based research leads to deeper understanding if investigators are forced to resolve "surprises," we describe next how variation in flow regimes reconfigured river food webs in the upper South Fork of the Eel River, as reported by Mary Power and her colleagues in 2008, and how these observations required us to rethink what we had learned from previous years of experiments and prevailing theory.

The Angelo Coast Range Reserve has protected about 5 km of the upper South Fork of the Eel River since the 1930s under the stewardship of Heath and Margorie Angelo, The Nature Conservancy (1959–1994), and the University of California Natural Reserve System (1994–present). Bill Trush, whose dissertation research at Angelo (Trush 1991) linked geomorphology and hydrology to steelhead biology, first told Mary Power about Angelo. Inspired by research at Sycamore Creek, Power's initial research at Angelo focused on nutrients, algae, and invertebrates. Fish seemed unimportant in the Eel, as they were neither as abundant nor as diverse as in rivers Power had studied previously in the midwestern United States and in Panama.

At the suggestion of Bill Dietrich (her geomorphologist husband), Power and her group began simple fish manipulations in 6 $m^2$ experimental pens in the summer of 1989. Surprisingly, predatory steelhead and adult roach had strong cascading effects on algae (Power 1990). In the presence of fish, attached green macroalgae (*Cladophora*) were nearly extirpated. In fish exclosures, *Cladophora* turfs remained 40–60 cm high and became overgrown by epiphytic diatoms (*Epithemia*) and free-living *Nostoc* cyanobacteria. Power would have thought *Epithemia* was "just another diatom," but Jane Marks and Rex Lowe pointed out its nitrogen-fixing endosymbionts, leading Angelo researchers to study its importance in Eel River food webs (Kupferberg et al. 1994; Power et al. 2009).

The negative effects of fish on algae were unexpected, because fish stomachs contained mostly algivorous invertebrates (mayflies and midges). One midge grazer, however, *Pseudochironomus richardsoni*, increased in fish enclosures, where it mowed down *Cladophora*, weaving it into knotty retreats (tufts). Fish had devoured *Pseudochironomus* extracted from tufts and tied on threads to assay predator risk across river habitats. But fish did not recognize midges inside tufts as prey, whereas small predators (e.g., lestid damselfly nymphs) did (Power et al. 1992). Enclosed fish consumed those small predators, resulting in a four-level food chain that suppressed algae. Small predators became abundant in fish exclosures, consuming all algivores, including *Pseudochironomus* (Power 1990). Released algae became resource limited (consistent with predictions by Hairston et al. 1960), explaining why nitrogen fixers (*Epithemia* and *Nostoc*) dominated the base of these three-level food chains.

Tim Wootton and Michael Parker joined Power the next field season and doubled the number of experimental pens to learn more about the effects of roach minnows versus steelhead parr on food webs. To their surprise, in 1990 and 1991, fish had no effects on algae. Within three or four weeks, attached algae disappeared in all experimental pens, with or without fish. Wootton recalls a common refrain from Power: "It's different this year." The researchers eventually recognized that drought had eliminated winter flood scour, so higher densities of algae-grazing armored caddisflies (*Dicosmoecus gilvipes*) had overwintered. By early summer, armored *Dicosmoecus* instars, which are invulnerable to most predators, were 1–2 orders of magnitude denser than during the post-flood summer. When *Dicosmoecus* were excluded from stream cages, algae recovered. Enclosed steelhead parr still had negative but relatively small effects on algae, revealing that top-down control ran primarily through a two-level food chain topped by predator-resistant grazers (Wootton et al. 1996).

Five years of repeated fish manipulations by Power and her colleagues showed that predator controls over summer algal biomass were strong following scouring winter floods and weak otherwise, suggesting that disturbances extirpating predator-resistant grazers were required for longer food chains. Previous theory held that disturbance should shorten food chains (May 1973; Pimm and Kitching 1987). But chains lengthen if most extirpated biomass is at lower trophic levels and if it is replaced by more vulnerable prey or producers, as commonly occurs during early succession (Power et al. 1996).

These experiments did not rule out bottom-up effects of floods on Eel River food webs. In addition to extirpating invulnerable grazers (top-down control), wet winters could improve summer growth conditions by delivering more nutrients or prolonging baseflows favorable to algal growth (bottom-up control). This prolonged place-based research program also resulted in opportunistic support for top-down control of algae mediated by floods. In winter 2009–2010, an unusually small (in areal extent) rainstorm confined to the catchment of one tributary, Elder Creek, elevated that stream's discharge well above scouring bankfull level, while discharge in the mainstem South Fork Eel remained below half bankfull. *Dicosmoecus* were exported from Elder Creek, which, during the following summer, developed the lushest standing crops of *Cladophora* ever observed there. *Dicosmoecus* remained abundant in the unscoured Eel, and summer *Cladophora* never proliferated. This unusual hydrologic event revealed that spatial contrasts in the mortality of predator-resistant grazers could reverse the gradient of algal biomass typically driven by down-channel increases in light (Finlay et al. 2011).

More evidence of top-down control of algae came from collaborations with sedimentologists. Dietrich told Power of work by Charles Nittrouer in marine canyons off the mouth of the Eel, which trap about 50% of the river's sediment export (Sommerfield et al. 2007). Nittrouer and colleagues had cored sediment deposits in those canyons and discovered records with layers dated with near-annual resolution back to 1918. Jack Sculley counted the freshwater epiphytic diatom frustules in those cores and found that high frustule counts corresponded to big *Cladophora*-epiphyte years in the Eel. In contrast, annual correlations between freshwater frustule densities and total precipitation were noisy and, surprisingly, slightly negative (Sculley et al. 2017). Those results suggested that over eighty-three years, bottom-up effects of floods had less impact on annual algal biomass than did extirpation of invulnerable grazers by flood scour. Thus, the frustration and distress of failing to repeat experimental results ultimately yielded an expanded understanding of food webs reset by disturbance. This work required year-to-year results from repeated experiments on food webs in the same landscape, with the same food-web members, but under different hydrologic regimes—all made possible by a protected research reserve.

Knowledge of river ecology at Angelo has been enriched by the research of many other individuals (e.g., Camille McNeely, Gilbert Cabana, Bill Rainey, John Sabo, Hiromi Uno, Justin Bastow, Charlene Ng, Paula Furey, Gabe Rossi, Suzanne Kelson, Phil Georgakakos, Daniella Rempe, Jesse Hahm, and David Dralle; see https://angelo.berkeley.edu/research). Space precludes telling their stories, which also entail the confronting of challenging surprises over the years to learn more.

Coweeta Hydrologic Laboratory: Wallace, J. B., S. L. Eggert, J. L. Meyer, and J. R. Webster. 2015. "Stream Invertebrate Productivity Linked to Forest Subsidies: 37 Stream-Years of Reference and Experimental Data." *Ecology* 96:1213–28.

Researchers at Coweeta Hydrologic Laboratory have built on extensive knowledge of land use and natural history at this steep, deeply shaded, forested landscape in the southern Appalachians, where terrestrial leaf litter is the primary source of organic matter to streams. Established in 1934 as a US Forest Service hydrologic laboratory, Coweeta's research mission expanded with the arrival of Bruce Wallace and Jack Webster at a partner institution, the University of Georgia. Over four subsequent decades of field experiments, Wallace and colleagues focused on how macroinvertebrates affect forest streams (Cuffney et al. 1990) and how forest-stream linkages affect stream animals (Wallace et al. 1997, 1999, 2015).

Bruce Wallace, Sue Eggert, Judy Meyer, and Jack Webster compiled an overview, published in 2015, of thirty-seven stream-years of experiments documenting litter effects on stream food webs to ask how forest inputs (leaf litter and wood) affect forested headwater streams. Their reference data from 1984 to 2006 encompassed the second-driest and second-wettest years of Coweeta's seventy-year precipitation record. The idea of a litter-exclusion study arose during a graduate course in stream ecology co-taught by Wallace and Meyer. Students read a review paper claiming no evidence that stream insects were limited by detritus, prompting Wallace to state, "I did not believe this for a second." Decades later, studies have demonstrated both energy and nutrient limitation of macroinvertebrate communities (Wallace et al. 1997; Cross et al. 2006; Demi et al. 2018) and even of salamander top predators (Johnson and Wallace 2005; Bumpers et al. 2015) in Coweeta's small, shaded headwater streams.

Wallace and colleagues excluded riparian inputs from 170 m of the treatment stream with a mesh canopy and lateral fence for thirteen years (1993–2006). From 1996 to 1998, they also removed all small wood from this stream; from 1998 to 2000, they removed all large wood; in 2000–2001, they added PVC pipes and plastic tubing to mimic wood structure; and finally, they added "fast" (labile) leaves (2001–2003), "slow" (refractory) leaves (2003–2005), and a mixture of both (2005–2006) to the stream. They compared changes in the litter-excluded and leaf-enriched streams with those in similarly sized, unmanipulated reference streams.

After three years of litter exclusion, Wallace and his colleagues observed dramatic losses of organic matter from depositional areas (gravel, pebble, cobble, and silt substrates), but little or no treatment effect in steeper reaches with unretentive bedrock substrates, where little storage of organic matter occurred naturally. Stores of organic matter in depositional areas remained near zero for eighty-five months. The research team was surprised that the effects of litter removal were so strong that densities of invertebrates in depositional substrates dropped to the level of those in bedrock substrates, overriding the long-observed spatial contrasts between depositional and erosional areas in streams throughout this region. Within a year of additions of labile, then refractory, then mixed leaves, invertebrate densities in depositional areas recovered to reference levels.

Prolonged investigations at Coweeta reinforced the initial finding of strong bottom-up effects of terrestrial litter on aquatic animals in forested headwater streams (Wallace et al. 1997, 2015). Abundance and biomass of most animal species (except grazers) decreased when detritus was excluded—and the bottom-up effect of organic matter propagated from detritivores to predators. Predator-prey coupling tightened as detrital resources dwindled, possibly due to loss of prey refuges as well as food. A particularly informative measure of invertebrate responses to both bottom-up and top-down food-web forcing was habitat-weighted invertebrate secondary production ($g/m^2/year$; see chapters 1 and 5). The linear relationship between predator and prey secondary production was strong and similar across streams over thirty-seven stream-years (slope of regression ~0.29), approximating published estimates of bioenergetic efficiencies for invertebrate predators. Remarkably, a quantitative relationship between litter inputs and invertebrate primary consumers similar to that documented at Coweeta was observed after a maize litter addition in Bluff River Cave, Alabama, where in situ primary production was nonexistent (Venarsky et al. 2018), further demonstrating the fundamental nature of this energetic linkage.

The Coweeta study was "one of the iconic experiments in ecology, because no one has conducted an ecosystem-level detritus manipulation experiment nor followed it for so long" (Robert Hall, personal communication). Furthermore, this decades-long sequence of experiments under the mentorship of Coweeta senior researchers inspired graduate students and postdocs from many institutions to build on the work that had gone before (e.g., Jon Benstead, Phillip Bumpers, Wyatt Cross, John Davis, Mick Demi, Sally Entrekin, Jennifer Greenwood, Vlad Gulis, Brent Johnson, John Kominoski, David Manning, and Cindy Tant).

As in other place-based programs, researchers at Coweeta experienced surprises. For example, long-term (five-year) nutrient addition to Coweeta streams shortened food chains, contrary to established theory (Oksanen et al. 1981; Pimm and Kitching 1987). Further investigation revealed a potential mechanism for this effect. In fertilized Coweeta streams, large, invulnerable primary consumers (*Pycnopsyche* caddisflies) dominated invertebrate guilds, supplanting smaller, more vulnerable prey taxa. As a result, energy flow to predators decreased (Davis et al. 2010), decoupling the strong correlation between primary consumer and predator production previously observed in unfertilized streams, as Wallace and his colleagues reported. These results, which took four to five years to manifest, suggest that increased productivity, if diverted to lower trophic levels, can shorten food chains by accelerating prey or producer escape (in size or numbers) from consumer control (Davis et al. 2010). According to Wallace, "without annual production measurements for multiple years and streams, we would not have been able to diagnose the long-term relationships between annual total invertebrate production and annual predator production." Similarly, the multi-year comparisons of summer food webs at Angelo showed that disturbance lengthened food chains (**Power et al. 2008**), rather than shortening them as theory would predict—here, time for growth, rather than elevated nutrients, allowed defended taxa to dominate lower trophic levels.

Research at Coweeta demonstrates how place-based field experiments not only reveal counterintuitive processes, but also show where in the landscape those processes operate, and where they change, or even reverse direction. For example, Rosemond et al. (2015) showed that enrichment of headwater streams released detrital food webs from N and P limitation in the deeply shaded upper catchment by increasing respiration and export of particulate organic carbon, thereby reducing the residence time of terrestrial detritus by about 50%. This effect would be reversed downstream where channels are wide enough that nutrient stimulation of photosynthesis enhances carbon storage. Documenting this transition helps researchers determine where nutrient loading reverses its impact on stream carbon storage.

Overall, the decades of research in Coweeta have quantified how changes in aquatic-terrestrial linkages mediated by fluxes of catchment vegetation or nutrients can alter stream food webs and ecosystems. This foundation of understanding has helped society recognize and anticipate how changes in climate, forests, or chemical fluxes will alter future forest stream ecosystems.

Rocky Mountain Biological Laboratory: Peckarsky, B. L., A. C. Encalada, and A. R. McIntosh. 2011. "Why Do Vulnerable Mayflies Thrive in Trout Streams?" *American Entomologist* 57:152–64.

Rocky Mountain Biological Laboratory (RMBL) was founded in 1928, nearly a half century before Bobbi Peckarsky started working there in 1974. Influenced by mentors Stanley Dodson and Dave Allan, Peckarsky sought to understand the role of predation in open systems, and RMBL's cold, fast-flowing, rocky-bottomed streams with snowmelt hydrology, characteristic of high elevations in the Rocky Mountains, fit the bill. Early research focused narrowly on interactions between predatory stoneflies and mayfly prey (**Peckarsky 1980** [chapter 4]; Peckarsky and Penton 1989). Subsequently, long-term, place-based research at RMBL in collaboration with colleagues (e.g., Marge Penton, Angus McIntosh, Brad Taylor, Billie Kerans, Maruxa Alvarez, Cathy Cowan, Jonas Dahl) and students (e.g., Pete Ode, Chester Anderson, Matt Harper, Matt Fuller, Jen Moslemi, Andrea Encalada, Kara Cromwell) has expanded our understanding of predation in stream food webs to encompass primary producers and top consumers (Peckarsky et al. 2010).

As in the other places featured in this chapter, many years of field observations at RMBL revealed surprises, including observations that ran counter to conventional ideas about predator-prey interactions, as Peckarsky, Encalada, and McIntosh reported in their 2011 paper. For example, *Baetis* mayflies, which are favored prey of drift-feeding brook trout (*Salvelinus fontinalis*), were more abundant in streams with trout populations than in fishless streams (Peckarsky et al. 2008), despite studies showing that both predation and predator-induced prey emigration produced the opposite effect. Incremental research, including detailed natural history studies, observation, demographic modeling, and phenomenological and mechanistic experimentation over twenty years, produced several new revelations to explain those unexpected observations. First, variation in prey abundance may be driven primarily by nonconsumptive effects, such as predator-induced changes in prey behavior and life-history strategies (McPeek and Peckarsky 1998). For example, effects of consumption by trout on prey abundance may be countered by (1) predator-induced reduction in prey emigration (Cooper et al. 1990) or (2) predator-induced acceleration of prey development to reduce duration of prey exposure to predators (Peckarsky et al. 2001, 2002). Therefore, the apparent anomaly of higher prey densities in streams where mortality from trout predation was greater (Peckarsky et al. 2008) could be explained by life-history impacts of predators that scare more prey than they eat.

Second, observations and experiments that tested for nonconsumptive effects of predators by dripping "essence of brook trout" into artificial or whole streams showed that probabilities of *Baetis* surviving the larval stage are equivalent in trout-occupied and fishless streams (Peckarsky et al. 2001, 2002), because trout-induced acceleration of larval development exposes *Baetis* to predation for shorter time periods. This equivalence, however, could not explain why *Baetis* are more abundant in trout streams than in fishless streams.

Third, Andrea Encalada observed that *Baetis* females favored oviposition sites under large rocks protruding from the water surface in fast-flowing locations (Encalada and Peckarsky 2006) and discovered that those substrates were more common in trout streams than in fishless streams (Encalada and Peckarsky 2011; see figure 21 in Peckarsky et al. 2011). Consequently, densities of *Baetis* eggs were higher in trout streams, and those higher densities were sustained throughout the larval stage (Encalada and Peckarsky 2011) because of equivalent probabilities of surviving the larval stage. This revelation illustrates that some mechanisms explaining unexpected variation in the abundance of favored prey of trout (e.g., selective oviposition of *Baetis*) may be independent of predation or predator density, and may instead depend on geomorphic and hydrologic attributes of streams that constrain successful recruitment (Peckarsky et al. 2000).

As has been pointed out previously (Hynes 1970; Holmes and Likens 2016), place-based research also fosters interdisciplinary collaborations among scientists. Andrew Wilcox, a hydro-geomorphologist, provided insights about how physical attributes of streams might explain spatial variation in *Baetis* abundance. Relationships between hydrologic parameters and *Baetis* drift probability did not support passive, abiotically driven drift; some observations were consistent with expectations (e.g., drift was positively correlated with Reynolds number), and others were surprising (e.g., drift was negatively related to the competence of streams to transport bed particles; Wilcox et al. 2008).

Continuity, collaboration, and communication among researchers, which are important components of successful place-based research programs, are facilitated at field stations, where immersion in science is the norm, with fewer distractions from home institutions. For example, continuity among generations of researchers in the rivers of RMBL linked initial studies of the effects of stonefly predation to subsequent studies on topics from grazer-algae interactions to fish-invertebrate-algae-nutrient interactions, to the discovery of nonconsumptive effects on life histories, to recruitment limitation, to effects of parasites on stream food webs, and ultimately to effects of disturbance on alternative community states and effects of climate change on stream communities (the latter topics are not covered in this review). Peckarsky's research group operates as a team: each person has their own project, yet collaborates on all other projects, making the fieldwork safer and more fun. Meetings and informal interactions in the field facilitate group planning of research protocols, discussions of problems, data analyses, and interpretation. All team members, regardless of academic status or experience, have significant input, and their ideas and perspectives influence and enrich concurrent projects.

Good research raises more questions than it answers. After fifty years of place-based research at RMBL, some puzzling observations remain unexplained. For example, what are the consequences for regional populations of the nonconsumptive effects that increase the numbers of adult *Baetis* dispersing from trout versus non-trout streams, but produce smaller females with reduced fecundities (Peckarsky et al. 2001)? Similarly, the cascading effects of mermithid parasites of *Baetis* on algal resources and on predation by brook trout (Cromwell 2018) suggest that parasite loads of *Baetis* may need to be taken into account in future studies of RMBL food webs. Finally, catchment-scale comparisons are needed to learn how climate change–induced alterations in hydrologic and temperature regimes of RMBL streams affect oviposition (Encalada and Peckarsky 2006), larval growth rates (Peckarsky et al. 2001), and emergence of invertebrates (Harper and Peckarsky 2006). With climate and other global changes, other surprises will arise; for example, the recent proliferation of a nuisance diatom at RMBL, *Didymosphenia geminata*, that over decades of drought has potentially regime-changing effects on benthic communities (Brogan et al. 2024). Long-term place-based research will help stream ecologists predict the consequences of altered snowmelt regimes and drying of RMBL streams that have the potential to cause shifts in key components of stream food webs.

## LOOKING FORWARD

We contend that ecology of place has had a disproportionate impact on the foundations of stream ecology, because long-term presence of researchers promotes shared awareness of natural history, appreciation of the more subtle aspects of nature, and consequently, a more thorough understanding of patterns and processes shaping stream ecosystems. Cumulative experiences of researchers over the long term at the five highlighted sites, as well as many other long-term field sites (see footnote), illustrate the value of place-based research and suggest that foundational understanding of stream ecology often arises from unexpected results, if those results trigger subsequent investigations. Determining causality is often easier in long-term, place-based research, which has some of the advantages of a well-controlled experiment—only one or a few possible ecosystem controls may have changed when subsequent outcomes differ from those previously observed. Doing research within the same general geological setting and regional climate also enables researchers to isolate shifts in process controls over space, such as the release of attached algae from light limitation (Finlay et al. 2011). Those fundamental processes and their controls are not unique to streams but operate across a wide range of ecosystems throughout the world.

Studies featured in this chapter demonstrate that in streams as in other ecosystems (e.g., Paine 1969), long-term place-based findings can challenge long-standing ecological theory. Long-term immersion in the natural history and ecology of a place enables researchers to witness over time the coevolution of changing environments, landscape contexts, and climate on ecosystem processes. To interpret those observations usefully, place-based research should be driven by hypotheses about causality. Big data derived from hypothesis-free monitoring can be repurposed to test hypotheses but may often be "mismatched to what we need to know" (Lindemayer and Likens 2018) to understand and forecast change. The process-based understanding needed for prediction may elude correlational studies if run by teams unfamiliar with local phenology or natural history (Sagoff 2019; Lindenmayer and Likens 2018).

We argue that understanding of ecosystems obtained through long-term question-driven research at one place need not lead to the parochial view that "my catchment is different." Rather, fundamental understanding of causal processes from one ecosystem is transportable to others, either to test its generality or learn why, in different contexts, different functional rates or rules apply. Cross-system comparisons complement or test generalities derived from one place, but such comparisons should be grounded in the thorough understanding of local natural history that is common to place-based research (Downes et al. 2021).

We hope our overviews of contributions of place-based research to the foundations of stream ecology will encourage young stream ecologists to commit to long-term field studies, and, importantly, strengthen interest in protecting ecosystems at catchment scales for long-term research and teaching. Those privileged to work at protected natural research reserves (Norris 1998) know they enable projects not often possible in other natural areas, such as national, state, or public parks, or reserves that prioritize conservation over research, where aesthetic considerations often rule out field manipulations, or where multiple stakeholders and changing management can derail long-term investigations. Field observatories and field stations must be protected and sustained for research, teaching, and outreach if we are to grasp and manage the environmental, ecological, and evolutionary changes now occurring at unprecedented rates in our world (Schubel et al. 2014).

# EPILOGUE

For the intrepid reader that made it this far—congratulations! We hope you've found this book useful and enjoyable. In closing, we would like to offer a few reflections.

First, it may be noticeable—and perhaps surprising—to some readers that this book lacks a chapter explicitly focused on anthropogenic disturbance or global change. Our decision not to include such a chapter emerged from the realization that the theme of understanding human activities, and their effects on running waters, has been threaded throughout the entire history of our discipline, and glimpses of this can be found in most chapters. For example, Ruth Patrick's foundational work on bioassessment came from growing concerns about sewage and industrial wastes. Noel Hynes's textbook *The Ecology of Running Waters* was published ten years *after* his opus, *The Biology of Polluted Waters*, a separate book published at a time when our understanding of—and ability to mitigate—pollution in freshwater was much more limited. We hope that readers recognize that the quest to understand how streams and rivers work has always been guided and motivated by an awareness of anthropogenic change and the desire to conserve these ecosystems.

We wish to make a second, related point. In assembling this compilation of papers, the editors were reminded of the rich and diverse set of ecological questions that have been addressed using streams as model systems. Thus, while there are clearly pressing reasons to study streams and rivers in the contexts of environmental degradation and global change, we also hope this book inspires a new generation to embrace the kind of curiosity and theory-driven science that has formed the bedrock of our current understanding. Indeed, there is a still a wealth of basic questions—from the organismal to ecosystem levels—waiting to be answered in these dynamic systems.

Third, when all the chapters are considered together, it is clear that new tools and techniques have both advanced and limited our understanding of streams. In the earliest days of stream ecology, the development of basic sampling equipment advanced our understanding of unique habitats (e.g., the hyporheic zone) and processes (e.g., invertebrate drift). Now, machine-learning approaches promise to automate sample processing and revolutionize the extraction of information from data gathered using automated sensors, drones, and satellites. While these tools and technologies unquestionably help generate new knowledge, they may also limit or unintentionally direct the focus of our questions and pursuits. For example, the application of stable isotopes in ecology accelerated our understanding of food webs and nutrient cycling, but diverted most of our focus to carbon and nitrogen, for which various stable isotopes that can be experimentally manipulated exist in detectable amounts. At the same time, our understanding of the cycling of phosphorus and other trace elements that lack easily measured or nonradioactive isotopes has lagged behind. Thus, as we develop new approaches and techniques for studying streams, we must ensure they do not deter or distract us from the important questions about the organisms, processes, and habitats that make these ecosystems so interesting, and that must be confronted to understand and protect stream and river ecosystems.

Finally, readers may have noticed that almost every foundational paper highlighted in this volume comes from studies in North America, Europe, Australia, and New Zealand, reflecting a clear imbalance in the dissemination of insights gathered from streams in other regions. Although we are heartened by an increasingly diverse set of contributions from other countries, there are still large geographic and social inequities in how research is supported financially and intellectually. Concerted efforts are needed to support, synthesize, and incorporate insights from all regions of the world, including ecological understanding derived from traditional or local knowledge that may not be published in common outlets. Such efforts are particularly important as many regions face increasing population growth and thus increasing demand for water, hydropower, and food resources provisioned by river ecosystems. The ability to respond to these challenges will only come from collective knowledge and partnerships that span broad geographic and societal domains.

# ACKNOWLEDGMENTS

The editors would like to acknowledge the many chapter authors and reviewers who contributed their time and energy to this project, as well as colleagues that supported us at early stages of preparation. The book benefited greatly from such a wide range of expertise and perspectives. We would also like to thank Joe Calamia, Executive Editor for Science and Technology at the University of Chicago Press, for his encouragement and dedication, as well as Matt Lang, Norma Sims Roche, and Nicholas Lilly for expert assistance during the editing and production phase. We are grateful for the support of our respective universities during the preparation of the manuscript. Finally, we wish to thank our graduate students as the primary inspiration for the book, and for the many lively discussions about foundational papers over the years.

# LITERATURE CITED

Abelho, M. 2001. "From Litterfall to Breakdown in Streams: A Review." *Scientific World Journal* 1:656–80.

Abelho, M. 2020. "ATP as a Measure of Microbial Biomass." In *Methods to Study Litter Decomposition, A Practical Guide*, 2nd ed., edited by F. Bärlocher, M. O. Gessner, and M. A. S. Graça, 291–99. Cham, Switzerland: Springer.

Acreman, M. C., and M. J. Dunbar. 2004. "Defining Environmental River Flow Requirements: A Review." *Hydrology and Earth System Sciences* 8:861–76.

Acreman, M. C., and A. J. D. Ferguson. 2010. "Environmental Flows and the European Water Framework Directive." *Freshwater Biology* 55:32–48.

Acuña, V., C. Vilches, and A. Giorgi. 2011. "As Productive and Slow as a Stream Can Be: The Metabolism of a Pampean Stream." *Journal of the North American Benthological Society* 30:71–83.

Alexander, R. B., R. A. Smith, and G. E. Schwarz. 2000. "Effect of Stream Channel Size on the Delivery of Nitrogen to the Gulf of Mexico." *Nature* 403:758–61.

Allan, J. D. 1978. "Trout Predation and the Size Composition of Stream Drift." *Limnology and Oceanography* 23:1231–37.

Allan, J. D. 1982. "The Effects of a Reduction in Trout Density on the Invertebrate Community of a Mountain Stream." *Ecology* 63:1444–55.

Allan, J. D. 2004. "Landscapes and Riverscapes: The Influence of Land Use on Stream Ecosystems." *Annual Review of Ecology, Evolution, and Systematics* 35:257–84.

Allan, J. D., and B. P. Feifarek. 1989. "Distances Travelled by Drifting Mayfly Nymphs: Factors Influencing Return to the Substrate." *Journal of the North American Benthological Society* 8:322–30.

Allen, D. C., D. A. Kopp, K. H. Costigan, T. Datry, B. Hugueny, D. S. Turner, G. S. Bodner, and T. J. Flood. 2019. "Citizen Scientists Document Long-Term

Streamflow Declines in Intermittent Rivers of the Desert Southwest, USA." *Freshwater Science* 38:244–56.

Allen, K. R. 1951. *The Horokiwi Stream, a Study of a Trout Population*. New Zealand Department of Fisheries Bulletin no. 10. Wellington, New Zealand: New Zealand Marine Dept.

Amoros, C., and A. L. Roux. 1988. "Interaction between Water Bodies within the Floodplains of Large Rivers: Function and Development of Connectivity." *Münstersche Geographische Arbeiten* 29:125–30.

Anderson, K. E., R. M. Nisbet, S. Diehl, and S. D. Cooper. 2005. "Scaling Population Responses to Spatial Environmental Variability in Advection-Dominated Systems." *Ecology Letters* 8:933–43.

Anderson, N. H., and D. M. Lehmkuhl. 1968. "Catastrophic Drift in a Woodland Stream." *Ecology* 49:198–206.

Anderson, N. H., and J. R. Sedell. 1979. "Detritus Processing by Macroinvertebrates in Stream Ecosystems." *Annual Review of Entomology* 24:351–78.

Anholt, B. R. 1995. "Density Dependence Resolves the Stream Drift Paradox." *Ecology* 76:2235–39.

Appling, A. P., R. O. Hall Jr., C. B. Yackulic, and M. Arroita. 2018. "Overcoming Equifinality: Leveraging Long Time Series for Stream Metabolism Estimation." *Journal of Geophysical Research: Biogeosciences* 123:624–45.

Appling, A. P., J. S. Read, L. A. Winslow, M. Arroita, E. S. Bernhardt, N. A. Griffiths, R. O. Hall Jr. et al. 2018. "The Metabolic Regimes of 356 Rivers in the United States." *Scientific Data* 5:180292.

Ardón, M., L. H. Zeglin, R. M. Utz, S. D. Cooper, W. K. Dodds, R. J. Bixby, A. S. Burdett et al. 2021. "Experimental Nitrogen and Phosphorus Enrichment Stimulates Multiple Trophic Levels of Algal and Detrital-Based Food Webs: A Global Meta-analysis from Streams and Rivers." *Biological Reviews* 96:692–715.

Armitage, K. B. 1958. "Ecology of Riffle Insects of the Firehole River, Wyoming." *Ecology* 39:571–80.

Armstrong, J. B., and D. E. Schindler. 2013. "Going with the Flow: Spatial Distributions of Juvenile Coho Salmon Track an Annually Shifting Mosaic of Water Temperature." *Ecosystems* 16:1429–41.

Arroita, M., A. Elosegi, and R. O. Hall Jr. 2019. "Twenty Years of Daily Metabolism Show Riverine Recovery following Sewage Abatement: Long-Term Recovery of a Polluted River." *Limnology and Oceanography* 64:S77–S92.

Arscott, D. B., K. Tockner, and J. V. Ward. 2005. "Lateral Organization of Aquatic Invertebrates along the Corridor of a Braided Floodplain River." *Journal of the North American Benthological Society* 24:934–54.

Arsuffi, T. L., and K. Suberkropp. 1985. "Selective Feeding by Stream Caddisfly (Trichoptera) Detritivores on Leaves with Fungal-Colonized Patches." *Oikos* 45:50–58.

Arthington, A. H., R. J. Naiman, M. E. McClain, and C. Nilsson. 2010. "Preserving the Biodiversity and Ecological Services of Rivers: New Challenges and Research Opportunities." *Freshwater Biology* 55:1–16.

Badcock, R. M. 1949. "Studies on Stream Life in Tributaries of the Welsh Dee." *Journal of Animal Ecology* 18:193–208.

Bailey, R. G. 1966. "Observations on the Nature and Importance of Organic Drift in a Devon River." *Hydrobiologia* 27:353–67.

Bain, M. B., J. T. Finn, and H. E. Booke. 1988. "Streamflow Regulation and Fish Community Structure." *Ecology* 69:382–92.

Baker, M. A., C. N. Dahm, and H. M. Valett. 1999. "Acetate Retention and Metabolism in the Hyporheic Zone of a Mountain Stream." *Limnology and Oceanography* 44:1530–39.

Baldocchi, D. D. 2020. "How Eddy Covariance Flux Measurements Have Contributed to Our Understanding of Global Change Biology." *Global Change Biology* 26:242–60.

Baldy, V, V. Gobert, F. Guerold, E. Chauvet, D. Lambrigot, and J.-Y. Charcosset. 2007. "Leaf Litter Breakdown Budgets in Streams of Various Trophic Status: Effects of Dissolved Inorganic Nutrients on Microorganisms and Invertebrates." *Freshwater Biology* 52:1322–35.

Ball, R. C., and F. F. Hooper. 1963. "Translocation of Phosphorus in a Trout Stream Ecosystem." In *Radioecology*, edited by V. Schultz and A. W. J. Klement, 217–28. Washington, DC: Reinhold Publishing Corporation and the American Institute of Biological Sciences.

Bärlocher, F. 1982. "Conidium Production from Leaves and Needles in Four Streams." *Canadian Journal of Botany* 60:1487–94.

Bärlocher, F., ed. 1992. *The Ecology of Aquatic Hyphomycetes*. Ecological Studies, vol. 94. Berlin: Springer-Verlag.

Bärlocher, F., M. O. Gessner, and M. A. S. Graça, eds. 2020. *Methods to Study Litter Decomposition: A Practical Guide*. 2nd ed. Cham, Switzerland: Springer-Nature Switzerland.

Bärlocher, F., and B. Kendrick. 1973. "Fungi and Food Preferences of *Gammarus pseudolimnaeus*." *Archiv für Hydrobiologie* 72:501–16.

Bärlocher, F., and B. Kendrick. 1974. "Dynamics of the Fungal Population on Leaves in a Stream." *Journal of Ecology* 62:761–91.

Barnes, J. R., and G. W. Minshall, eds. 1983. *Stream Ecology: Application and Testing of General Ecological Theory*. New York: Plenum Press.

Bartley, T. J., K. S. McCann, C. Bieg, K. Cazelles, M. Granados, M. M. Guzzo, A. S. MacDougall, T. D. Tunney, and B. C. McMeans. 2019. "Food Web Rewiring in a Changing World." *Nature Ecology and Evolution* 3:345–54.

Baruch, E. M., T. K. Harms, A. Ruhi, M. Lu, L. Gaines-Sewell, and J. L. Sabo. 2023. "Food Web Efficiency in Desert Streams." *Limnology and Oceanography* 68:723–34.

Battin, T. J., L. A. Kaplan, S. Findlay, C. S. Hopkinson, E. Martí, A. I. Packman, J. D. Newbold, and F. Sabater. 2008. "Biophysical Controls on Organic Carbon Fluxes in Fluvial Networks." *Nature Geoscience* 1:95–100.

Battin, T. J., R. Lauerwald, E. S. Bernhardt, E. Bertuzzo, L. G. Gener, R. O. Hall Jr., E. R. Hotchkiss et al. 2023. "River Ecosystem Metabolism and Carbon Biogeochemistry in a Changing World." *Nature* 613:449–59.

Baxter, C. V., K. D. Fausch, M. Murakami, and P. L. Chapman, P. L. 2004. "Fish Invasion Restructures Stream and Forest Food Webs by Interrupting Reciprocal Prey Subsidies." *Ecology* 85:2656–63.

Baxter, C. V., K. D. Fausch, and W. C. Saunders. 2005. "Tangled Webs: Reciprocal Flows of Invertebrate Prey Link Streams and Riparian Zones." *Freshwater Biology* 50:201–20.

Baynes, E. R., M. Attal, S. Niedermann, L. A. Kirstein, A. J. Dugmore, and M. Naylor. 2015. "Erosion during Extreme Flood Events Dominates Holocene Canyon Evolution in Northeast Iceland." *Proceedings of the National Academy of Sciences* 112:2355–60.

Beaulieu, J. J., J. L. Tank, S. K. Hamilton, W. M. Wollheim, R. O. Hall Jr., P. J. Mulholland, B. J. Peterson et al. 2011. "Nitrous Oxide Emission from Denitrification in Stream and River Networks." *Proceedings of the National Academy of Sciences* 108:214–19.

Beechie, T. J., D. A. Sear, J. D. Olden, G. R. Pess, J. M. Buffington, H. Moir, P. Roni, and M. M. Pollock. 2010. "Process-Based Principles for Restoring River Ecosystems." *BioScience* 60:209–22.

Bellmore, J. R., and C. V. Baxter. 2014. "Effects of Geomorphic Process Domains on River Ecosystems: A Comparison of Floodplain and Confined Valley Segments." *River Research and Applications* 30:617–30.

Benda, L. E. E., N. L. Poff, D. Miller, T. Dunne, G. Reeves, G. Pess, and M. Pollock. 2004. "The Network Dynamics Hypothesis: How Channel Networks Structure Riverine Habitats." *BioScience* 54:413–27.

Ben-David, M., T. A. Hanley, and D. M. Schell. 1998. "Fertilization of Terrestrial Vegetation by Spawning Pacific Salmon: The Role of Flooding and Predator Activity." *Oikos* 83:47–55.

Benke, A. C. 2010. "Secondary Production as Part of Bioenergetic Theory: Contributions from Freshwater Benthic Science." *River Research and Applications* 26:36–44.

Benke, A. C. 2018. "River Food Webs: An Integrative Approach to Bottom-Up Flow Webs and Top-Down Impact Webs, and Trophic Position." *Ecology* 99:1370–81.

Benke, A. C., and A. D. Huryn. 2017. "Secondary Production and Quantitative Food Webs." In *Methods in Stream Ecology*, vol. 2, edited by G. A. Lamberti and F. R. Hauer, 235–54. Cambridge, MA: Academic Press, Elsevier.

Benke, A. C., and D. I. Jacobi. 1994. "Production Dynamics and Resource Utilization of Snag-Dwelling Mayflies in a Blackwater River." *Ecology* 75:1219–32.

Benke, A. C., and J. B. Wallace. 1980. "Trophic Basis for Production among Net-Spinning Caddisflies in a Southern Appalachian Stream." *Ecology* 61:108–18.

Benstead, J. P., and D. S. Leigh. 2012. "An Expanded Role for River Networks." *Nature Geoscience* 5:678–79.

Berg, K. 1948. "Biological Studies of the River Susaa." *Folia Limnologica Scandinavica* 4:1–318.

Berner, L. M. 1951. "Limnology of the Lower Missouri River." *Ecology* 32:1–12.

Bernhardt, E. S., J. B. Heffernan, N. B. Grimm, E. H. Stanley, J. W. Harvey, M. Arroita,

A. P. Appling et al. 2018. "The Metabolic Regimes of Flowing Waters." *Limnology and Oceanography* 63:S99–S118.

Bernhardt, E. S., and G. E. Likens. 2004. "Controls on Periphyton Biomass in Heterotrophic Streams." *Freshwater Biology* 49:14–27.

Bernhardt, E. S., G. E. Likens, D. C. Buso, and C. T. Driscoll. 2003. "In-Stream Uptake Dampens Effects of Major Forest Disturbance on Watershed Nitrogen Export." *Proceedings of the National Academy of Sciences* 100:10304–8.

Bernhardt, E. S., G. E. Likens, R. O. Hall Jr., D. C. Buso, S. G. Fisher, T. M. Burton, J. L. Meyer et al. 2005. "Can't See the Forest for the Stream? The Capacity of In-Stream Processing to Modify Terrestrial Nitrogen Exports." *BioScience* 55:219–30.

Bernhardt, E. S., M. A. Palmer, J. D. Allan, G. Alexander, K. Barnas, S. Brooks, J. Carr et al. 2005. "Synthesizing US River Restoration Efforts." *Science* 308:636–37.

Bernhardt, E. S., P. Savoy, M. J. Vlah, A. P. Appling, L. E. Koenig, R. O. Hall Jr., M. Arroita et al. 2022. "Light and Flow Regimes Regulate the Metabolism of Rivers." *Proceedings of the National Academy of Sciences* 119: e2121976119.

Bilby, R. E. 1981. "Role of Organic Debris Dams in Regulating the Export of Dissolved and Particulate Matter from a Forested Watershed." *Ecology* 62:1234–43.

Bilby, R. E., B. R. Fransen, and P. A. Bisson. 1996. "Incorporation of Nitrogen and Carbon from Spawning Coho Salmon into the Trophic System of Small Streams: Evidence from Stable Isotopes." *Canadian Journal of Fisheries and Aquatic Sciences* 53:164–73.

Billick, I. A., and M. V. Price, eds. 2010. *The Ecology of Place*. Chicago: University of Chicago Press.

Bird, G. A., and N. K. Kaushik. 1981. "Coarse Particulate Organic Matter in Streams." In *Perspectives in Running Water Ecology*, edited by M. A. Lock and D. D. Williams, 41–68. New York: Plenum Press.

Bisson, P. A., J. A. Nielsen, R. A. Palmason, and E. L. Grove. 1982. "A System of Naming Habitat Types in Small Streams, with Examples of Habitat Utilization by Salmonids during Low Stream Flow." In *Acquisition and Utilization of Aquatic Habitat Inventory Information*, edited by N. B. Armantrout, 62–73. Portland, OR: Western Division, American Fisheries Society.

Blöschl, G., J. Hall, J. Parajka, R. A. P. Perdigão, B. Merz, B. Arheimer, G. T. Aronica, A. Bilibashi, O. Bonacci, and M. Borga. 2017. "Changing Climate Shifts Timing of European Floods." *Science* 357:588–90.

Bonada, N., N. Prat, V. H. Resh, and B. Statzner. 2006. "Developments in Aquatic Insect Biomonitoring: A Comparative Analysis of Recent Approaches." *Annual Review of Entomology* 51:495–523.

Bormann, F. H., and G. E. Likens. 1967. "Nutrient Cycling." *Science* 155:424–29.

Boulton, A. J. 1989. "Over-summering Refuges of Aquatic Macroinvertebrates in Two Intermittent Streams in Central Victoria." *Transactions of the Royal Society of South Australia* 113:23–34.

Boulton, A. J., and P. I. Boon. 1991. "A Review of Methodology Used to Measure Leaf Litter Decomposition in Lotic Environments: Time to Turn Over an Old Leaf?" *Australian Journal of Marine and Freshwater Research* 42:1–43.

Boulton, A. J., S. Findlay, P. Marmonier, E. H. Stanley, and H. M. Valett. 1998. "The Functional Significance of the Hyporheic Zone in Streams and Rivers." *Annual Review of Ecology and Systematics* 29:59–81.

Boyd, M., J. T. Teller, Z. Yang, L. Kingsmill, and C. Shultis. 2012. "An 8,900-Year-Old Forest Drowned by Lake Superior: Hydrological and Paleoecological Implications." *Journal of Paleolimnology* 47:339–55.

Boyero, L., M. O., Gessner, R. G. Pearson, E. Chauvet, J. Pérez, S. D. Tiegs, A. M. Tonin, F. Correa-Araneda, N. López-Rojo, and M. A. Graça. 2021. "Global Patterns of Plant Litter Decomposition in Streams." In *The Ecology of Plant Litter Decomposition in Stream Ecosystems*, edited by C. M. Swan, L. Boyero, and C. Canhoto, 51–71. Cham, Switzerland: Springer-Nature Switzerland.

Boyero, L., N. López-Rojo, A. M. Tonin, J. Pérez, F. Correa-Araneda, R. G. Pearson, J. Bosch et al. 2021. "Impacts of Detritivore Diversity Loss on Instream Decomposition Are Greatest in the Tropics." *Nature Communications* 12:3700.

Boyero, L., R. G. Pearson, M. O. Gessner, L. A. Barmuta, V. Ferreira, M. A. S. Graça, D. Dudgeon et al. 2011. "A Global Experiment Suggests Climate Warming Will Not Accelerate Litter Decomposition in Streams but May Reduce Carbon Sequestration." *Ecology Letters* 14:289–94.

Boyero, L., J. Pérez, N. López-Rojo, A. M. Tonin, F. Correa-Araneda, R. G. Pearson, J. Bosch et al. 2021. "Latitude Dictates Plant Diversity Effects on Instream Decomposition." *Science Advances* 7:eabe7860.

Boyero, L., A. Ramírez, D. Dudgeon, and R. G. Pearson. 2009. "Are Tropical Streams Really Different?" *Journal of the North American Benthological Society* 28:397–403.

Brardinoni, F., and M. A. Hassan. 2006. "Glacial Erosion, Evolution of River Long Profiles, and the Organization of Process Domains in Mountain Drainage Basins of Coastal British Columbia." *Journal of Geophysical Research: Earth Surface* 111:F1.

Briand, F. 1983. "Environmental Control of Food Web Structure." *Ecology* 64:253–63.

Briand, F., and J. E. Cohen. 1987. "Environmental Correlates of Food Chain Length." *Science* 238:956–60.

Brierley, G. J., and K. Fryirs. 2000. "River Styles, a Geomorphic Approach to Catchment Characterization: Implications for River Rehabilitation in Bega Catchment, New South Wales, Australia." *Environmental Management* 25:661–79.

Briers, R. A., J. H. Gee, H. M. Cariss, and R. Geoghegan. 2004. "Inter-population Dispersal by Adult Stoneflies Detected by Stable Isotope Enrichment." *Freshwater Biology* 49:425–31.

Brighenti, S., S. Hotaling, D. S. Finn, A. G. Fountain, M. Hayashi, D. Herbst, J. E. Saros, L. M. Tronstad, and C. I. Millar. 2021. "Rock Glaciers and Related Cold Rocky Landforms: Overlooked Climate Refugia for Mountain Biodiversity." *Global Change Biology* 27:1504–17.

Brinkhurst, R. O. 1965. "Observations on the Recovery of a British River from Gross Organic Pollution." *Hydrobiologia* 25:9–51.

Brisbane Declaration. 2007. "Environmental Flows Are Essential for Freshwater Ecosystem Health and Human Well-Being." Declaration of the 10th

International River Symposium and International Environmental Flows Conference, 3–6 September 2007: Brisbane, Australia.

Brittain, J. E., and T. J. Eikeland. 1988. "Invertebrate Drift: A Review." *Hydrobiologia* 166:77–93.

Brogan, M., B. L. Peckarsky, and J. Resasco. 2024. "Consequences of Nuisance Algal Blooms of *Didymosphenia geminata* on Invertebrate Communities in Rocky Mountain Streams." *Freshwater Science* 43:52–64.

Brooks, J. L., and S. L. Dodson. 1965. "Predation, Body Size, and Composition of Plankton." *Science* 150:28–35.

Brookshire, E. N., H. M. Valett, and S. Gerber. 2009. "Maintenance of Terrestrial Nutrient Loss Signatures during In-Stream Transport." *Ecology* 90:293–99.

Brose, U., P. Archambault, A. D. Barnes, L. F. Bersier, T. Boy, J. Canning-Clode, E. Conti, M. Dias, C. Digel, A. Dissanayake, and A. A. Flores. 2019. "Predator Traits Determine Food-Web Architecture across Ecosystems." *Nature Ecology and Evolution* 3:919–27.

Brown, B. L., and C. M. Swan. 2010. "Dendritic Network Structure Constrains Metacommunity Properties in Riverine Ecosystems." *Journal of Animal Ecology* 79:571–80.

Brown, D. S. 1961. "The Food of the Larvae of *Cloeon dipterum* L. and *Baetis rhodani* (Pictet) (Insecta, Ephemeroptera)." *Journal of Animal Ecology* 30:55–75.

Bruns, D. A., G. W. Minshall, C. E. Cushing, K. W. Cummins, J. T. Brock, and R. L. Vannote. 1984. "Tributaries as Modifiers of the River Continuum Concept: Analysis by Polar Ordination and Regression Models." *Archiv für Hydrobiologie* 99:208–20.

Bumpers, P. M., J. C. Maerz, A. D. Rosemond, and J. P. Benstead. 2015. "Salamander Growth Rates Increase along an Experimental Phosphorus Gradient." *Ecology* 96:2994–3004.

Bumpers, P. M., A. D. Rosemond, J. C. Maerz, and J. P. Benstead. 2017. "Experimental Nutrient Enrichment of Forest Streams Increases Energy Flow to Predators along Greener Food-Web Pathways." *Freshwater Biology* 62:1794–1805.

Burchsted, D., M. Daniels, R. Thorson, and J. Vokoun. 2010. "The River Discontinuum: Applying Beaver Modifications to Baseline Conditions for Restoration of Forested Headwaters." *BioScience* 60:908–22.

Busch, D. E., and S. G. Fisher. 1981. "Metabolism of a Desert Stream." *Freshwater Biology* 11:301–7.

Campbell Grant, E. H., W. H. Lowe, and W. F. Fagan. 2007. "Living in the Branches: Population Dynamics and Ecological Processes in Dendritic Networks." *Ecology Letters* 10:165–75.

Cardenas, M. B., J. L. Wilson, and V. A. Zlotnik. 2004. "Impact of Heterogeneity, Bed Forms, and Stream Curvature on Subchannel Hyporheic Exchange." *Water Resources Research* 40:W08307.

Carpenter, K. E. 1927. "Faunistic Ecology of Some Cardiganshire Streams." *Journal of Ecology* 15:33–54.

Cederholm, C. J., M. D. Kunze, T. Murota, and A. Sibatani. 1999. "Pacific Salmon Carcasses: Essential Contributions of Nutrients and Energy for Aquatic and Terrestrial Ecosystems." *Fisheries* 24:6–15.

Chan, E. K., Y. Zhang, and D. Dudgeon. 2008. "Arthropod 'Rain' into Tropical Streams: The Importance of Intact Riparian Forest and Influences on Fish Diets." *Marine and Freshwater Research* 59:653–60.

Chapman, D. W. 1966. "The Relative Contributions of Aquatic and Terrestrial Primary Producers to the Trophic Relations of Stream Organisms." *Special Publications of the Pymatuning Laboratory of Field Biology* 4:116–30.

Chapman, D. W., and R. Demory. 1963. "Seasonal Changes in the Food Ingested by Aquatic Insect Larvae and Nymphs in Two Oregon Streams." *Ecology* 44:140–46.

Chauvet, E., V. Ferreira, P. S. Giller, B. G. McKie, S. D. Tiegs, G. Woodward, A. Elosegi et al. 2016. "Litter Decomposition as an Indicator of Stream Ecosystem Functioning at Local-to-Continental Scales: Insights from the European RivFunction Project." *Advances in Ecological Research* 55:99–182.

Clarke, A., R. M. Nally, N. Bond, and P. Lake. 2008. "Macroinvertebrate Diversity in Headwater Streams: A Review." *Freshwater Biology* 53:1707–21.

Closs, G. P., and P. S. Lake. 1994. "Spatial and Temporal Variation in the Structure of an Intermittent-Stream Food Web." *Ecological Monographs* 64:1–21.

Collins, B. D., D. R. Montgomery, K. L. Fetherston, and T. B. Abbe. 2012. "The Floodplain Large-Wood Cycle Hypothesis: A Mechanism for the Physical and Biotic Structuring of Temperate Forested Alluvial Valleys in the North Pacific Coastal Ecoregion." *Geomorphology* 139:460–70.

Collins, S. F., C. V. Baxter, A. M. Marcarelli, L. Felicetti, S. Florin, M. S. Wipfli, and G. Servheen. 2020. "Reverberating Effects of Resource Exchanges in Stream-Riparian Food Webs." *Oecologia* 192:179–89.

Collins, S. M., S. A. Thomas, T. Heatherly, K. L. MacNeill, A. O. Leduc, A. López-Sepulcre, B. A. Lamphere et al. 2016. "Fish Introductions and Light Modulate Food Web Fluxes in Tropical Streams: A Whole-Ecosystem Experimental Approach." *Ecology* 97:3154–66.

Connell, J. H. 1978. "Diversity on Tropical Rainforests and Coral Reefs." *Science* 199:1302–10.

Cooke, S. J., J. C. Vermaire, H. M. Baulch, K. Birnie-Gauvin, W. M. Twardek, and J. S. Richardson. 2022. "Our Failure to Protect the Stream and Its Valley: A Call to Back Off from Riparian Development." *Freshwater Science* 41:183–94.

Cooper, S. D., S. J. Walde, and B. L. Peckarsky. 1990. "Prey Exchange Rates and the Impact of Predators on Prey Populations in Streams." *Ecology* 71:1503–14.

Covino, T. P., B. L. McGlynn, and R. A. McNamara. 2010. "Tracer Additions for Spiraling Curve Characterization (TASCC): Quantifying Stream Nutrient Uptake Kinetics from Ambient to Saturation." *Limnology and Oceanography: Methods* 8:484–98.

Cromwell, K. 2018. "Host-Parasite Ecology of *Baetis bicaudatus* Mayflies and *Gasteromermis* spp.: Nematodes in High Altitude Streams." PhD dissertation, University of Wisconsin–Madison.

Cross, W. F., C. V. Baxter, E. J. Rosi-Marshall, R. O. Hall Jr., T. A. Kennedy, K. C. Donner, H. A. Wellard Kelly et al. 2013. "Food-Web Dynamics in a Large River Discontinuum." *Ecological Monographs* 83:311–37.

Cross, W. F., J. B. Wallace, A. D. Rosemond, and S. L. Eggert. 2006. "Whole-System Nutrient Enrichment Increases Secondary Production in a Detritus-Based Ecosystem." *Ecology* 87:1556–65.

Crowl, T. A., and A. P. Covich. 1990. "Predator-Induced Life-History Shifts in a Freshwater Snail." *Science* 247:949–51.

Crowl, T. A., C. R. Townsend, and A. R. McIntosh. 1992. "The Impact of Introduced Brown and Rainbow Trout on Native Fish: The Case of Australasia." *Reviews in Fish Biology and Fisheries* 2:217–41.

Cuffney, T. F., J. B. Wallace, and G. J. Lugthart. 1990. "Experimental Evidence Quantifying the Role of Benthic Invertebrates in Organic Matter Dynamics of Headwater Streams." *Freshwater Biology* 23:281–99.

Cummins, K. W. 1966. "A Review of Stream Ecology with Special Emphasis on Organism-Substrate Relationships." *Special Publications of the Pymatuning Laboratory of Field Biology* 4:2–51.

Cummins, K. W. 1973. "Trophic Relations of Aquatic Insects." *Annual Review of Entomology* 18:183–206.

Cummins, K. W. 1974. "Structure and Function of Stream Ecosystems." *BioScience* 24:631–41.

Cummins, K. W., W. P. Coffman, and P. A. Roff. 1966. "Trophic Relations in a Small Woodland Stream." *Internationalen Vereinigung für Theoretische und Angewandte Limnologie: Verhandlungen* 16:627–38.

Cummins, K. W., and G. H. Lauff. 1969. "The Influence of Substrate Particle Size on the Microdistribution of Stream Macrobenthos." *Hydrobiologia* 34:145–81.

Cummins, K. W., R. C. Petersen, F. O. Howard, J. C. Wuycheck, and V. I. Holt. 1973. "The Utilization of Leaf Litter by Stream Detritivores." *Ecology* 54:336–45.

Cummins, K. W., G. L. Spengler, G. M. Ward, R. M. Speaker, R. W. Ovink, D. C. Mahan, and R. L. Mattingly. 1980. "Processing of Confined and Naturally Entrained Leaf Litter in a Woodland Stream Ecosystem." *Limnology and Oceanography* 25:952–57.

Cummins, K. W., M. A. Wilzbach, D. M. Gates, J. B. Perry, and W. B. Taliaferro. 1989. "Shredders and Riparian Vegetation." *BioScience* 39:24–30.

Cushing, C. E., and E. G. Wolf. 1982. "Organic Energy Budget of Rattlesnake Springs, Washington." *American Midland Naturalist* 107:404–7.

Cushing, C. E., and E. G. Wolf. 1984. "Primary Production in Rattlesnake Springs, a Cold Desert Spring-Stream." *Hydrobiologia* 114:229–36.

Darwin, C. R. 1883. *The Formation of Vegetable Mould through the Action of Worms with Observations on Their Habits*. London: John Murray.

Datry, T., N. Bonada, and A. J. Boulton. 2017. *Intermittent Rivers and Ephemeral Streams: Ecology and Management*. Cambridge, MA: Academic Press, Elsevier.

Datry, T., A. Foulquier, R. Corti, D. von Schiller, K. Tockner, C. Mendoza-Lera, J. C. Clément et al. 2018. "A Global Analysis of Terrestrial Plant Litter Dynamics in Non-perennial Waterways." *Nature Geoscience* 11:497–503.

Datry, T., S. T. Larned, and K. Tockner. 2014. "Intermittent Rivers: A Challenge for Freshwater Ecology." *BioScience* 64:229–35.

Davidson, F. A., and J. L. Wilding. 1943. "A Quantitative Faunal Investigation of a Cold Spring Community." *American Midland Naturalist* 29:200–209.

Davis, C. J., C. H. Fritsen, E. D. Wirthlin, and J. C. Memmott. 2012. "High Rates of Primary Production in a Semi-arid Tailwater: Implications for Self-Regulated Production." *River Research and Applications* 28:1820–29.

Davis, J. M., A. D. Rosemond, S. L. Eggert, W. F. Cross, and J. B. Wallace. 2010. "Long-Term Nutrient Enrichment Decouples Predator and Prey Production." *Proceedings of the National Academy of Sciences* 107:121–26.

Death, R. G., and M. J. Winterbourn. 1995. "Diversity Patterns in Stream Benthic Invertebrate Communities: The Influence of habitat Stability." *Ecology* 76:1446–60.

Deiner, K., E. A. Fronhofer, E. Mächler, J. C. Walser, and F. Altermatt. 2016. "Environmental DNA Reveals That Rivers Are Conveyer Belts of Biodiversity Information." *Nature Communications* 7:12544.

Demi, L. M., J. P. Benstead, A. D. Rosemond, and J. C. Maerz. 2018. "Litter P Content Drives Consumer Production in Detritus-Based Streams Spanning an Experimental N:P Gradient." *Ecology* 99:347–59.

Dimond, J. B. 1967. "Evidence That Drift of Stream Benthos Is Density Related." *Ecology* 48:855–57.

Dodds, G. S., and F. L. Hisaw. 1925. "Ecological Studies of Aquatic Insects. IV. Altitudinal Range and Zonation of Mayflies, Stoneflies and Caddisflies in the Colorado Rockies." *Ecology* 6:380–90.

Dodds, W. K., K. Gido, M. R. Whiles, K. M. Fritz, and W. J. Matthews. 2004. "Life on the Edge: The Ecology of Great Plains Prairie Streams." *BioScience* 54:205–16.

Dodds, W. K., C. T. Robinson, E. Galser, G. J. Anderson Hansen, H. Powell, J. M. Smith, N. Morse et al. 2012. "Surprises and Insights from Long-Term Aquatic Data Sets and Experiments." *BioScience* 62:709–21.

Dodson, S. I., T. A. Crowl, B. L. Peckarsky, L. B. Kats, A. P. Covich, and J. M. Culp. 1994. "Non-visual Communication in Freshwater Benthos: An Overview." *Journal of the North American Benthological Society* 13:268–82.

Dole-Olivier, M. J., P. Marmonier, M. C. Des Chatelliers, and D. Martin. 1994. "Interstitial Fauna Associated with the Alluvial Floodplains of the Rhône River (France)." In *Groundwater Ecology*, edited by J. Gibert, D. L. Danielopol, and J. A. Stanford, 312–46. New York: Academic Press.

Downes, B. J., P. S. Lake, and E. S. G. Schreiber. 1993. "Spatial Variation in the Distribution of Stream Invertebrates: Implications of Patchiness for Models of Community Organization." *Freshwater Biology* 30:119–32.

Downes, B. J., and J. Lancaster. 2018. "Itinerant, Nomad or Invader? A Field Experiment Sheds Light on the Characteristics of Successful Dispersers and Colonists." *Freshwater Biology* 63:1394–1406.

Downes, B. J., B. L. Peckarsky, J. Lancaster, W. Boville, and M. Alp. 2021. "From Insects to Frogs, Egg-Juvenile Recruitment Can Have Persistent Effects on Population Sizes." *Annual Review of Ecology, Evolution and Systematics* 52:67–86.

Duda, J. J., C. E. Torgersen, S. J. Brenkman, R. J. Peters, K. T. Sutton, H. A. Connor, P. Kennedy et al. 2021. "Reconnecting the Elwha River: Spatial Patterns of Fish Response to Dam Removal." *Frontiers in Ecology and Evolution* 9:765488.

Dudgeon, D. 1982. "An Investigation of Physical and Biological Processing of Two Species of Leaf Litter in Tai Po Kau Forest Streams, New Territories, Hong Kong." *Archiv für Hydrobiologie* 96:1–32.

Dudgeon, D., A. H. Arthington, M. O. Gessner, Z.-I. Kawabata, D. J. Knowler, C. Lévêque, R. J. Naiman et al. 2006. "Freshwater Biodiversity: Importance, Threats, Status and Conservation Challenges." *Biological Reviews* 81:163–82.

Duffer, W., and T. C. Dorris. 1966. "Primary Productivity in a Southern Great Plains Stream." *Limnology and Oceanography* 11:143–51.

Edington, J. M. 1968. "Habitat Preferences in Net-Spinning Caddis Larvae with Special Reference to the Influence of Water Velocity." *Journal of Animal Ecology* 37:675–92.

Edwards, A. M. C. 1974. "Silicon Depletions in Some Norfolk Rivers." *Freshwater Biology* 4:267–74.

Egglishaw, H. J. 1964. "The Distributional Relationship between the Bottom Fauna and Plant Detritus in Streams." *Journal of Animal Ecology* 463–76.

Egglishaw, H. J. 1972. "Experimental Study of the Breakdown of Cellulose in Fast Flowing Streams." *Memorie Dell'istituto Italiano Di Idrobiologia* (Italy) 29:405–28.

Elliott, J. M. 1965. "Daily Fluctuations of Drift Invertebrates in a Dartmoor Stream." *Nature* 205:1127–29.

Elliot, J. M. 1967. "Invertebrate Drift in a Dartmoor Stream." *Archiv für Hydrobiologie* 63:202–37.

Elliott, J. M. 1970. "Diel Changes in Invertebrate Drift and the Food of Trout *Salmo trutta* L." *Journal of Fish Biology* 2:161–65.

Elliott, J. M. 1971. "The Distances Travelled by Drifting Invertebrates in a Lake District Stream." *Oecologia* 6:191–220.

Elliott, J. M. 1973. "The Food of Brown and Rainbow Trout (*Salmo trutta* and *S. gairdneri*) in Relation to the Abundance of Drifting Invertebrates in a Mountain Stream." *Oecologia* 12:329–47.

Ellis, M. M. 1937. "Detection and Measurement of Stream Pollution." US Bureau of Fisheries, Bulletin 22.

Ellis, R. J., and H. Gowing. 1957. "Relationship between Food Supply and Condition of Wild Brown Trout, *Salmo trutta* Linnaeus, in a Michigan Stream." *Limnology and Oceanography* 2:299–308.

Elton, C. S. 1927. *Animal Ecology*. London: Sedgwick & Jackson.

Elwood, J. W., and D. J. Nelson. 1972. "Periphyton Production and Grazing Rates in a Stream Measured with a $^{32}P$ Material Balance Method." *Oikos* 32:295–303.

Elwood, J. W., J. D. Newbold, R. V. O'Neill, and W. Van Winkle. 1983. "Resource Spiraling: An Operational Paradigm for Analyzing Lotic Ecosystems." In *Dynamics of Lotic Ecosystems*, edited by T. D. Fontaine, and S. M. Bartell, 3–27. Ann Arbor, MI: Ann Arbor Science.

Elwood, J. W., J. D. Newbold, A. F. Trimble, and R. W. Stark. 1981. "The Limiting Role of Phosphorus in a Woodland Stream Ecosystem: Effects of P-enrichment on Leaf Decomposition and Primary Producers." *Ecology* 62:146–58.

Encalada, A. C., and B. L. Peckarsky. 2006. "Selective Oviposition by the Mayfly *Baetis bicaudatus*." *Oecologia* 148:526–37.

Encalada, A. C., and B. L. Peckarsky. 2011. "The Influence of Recruitment on Within-Generation Population Dynamics of a Mayfly." *Ecosphere* 2:1–14.

Encalada, A. C., and B. L. Peckarsky. 2012. "Large-Scale Manipulation of Mayfly Recruitment Affects Population Size." *Oecologia* 168:967–76.

Englund, G. 1997. "Importance of Spatial Scale and Prey Movements in Predator Caging Experiments." *Ecology* 78:2316–25.

Ensign, S. H., M. W. Doyle, and J. R. Gardner. 2017. "New Strategies for Measuring Rates of Environmental Processes in Rivers, Lakes, and Estuaries." *Freshwater Science* 36:453–65.

Erdozain, M., K. Kidd, D. Kreutzweiser, and P. Sibley. 2019. "Increased Reliance of Stream Macroinvertebrates on Terrestrial Food Sources Linked to Forest Management Intensity." *Ecological Applications* 29:e01889.

Fausch, K. D. 1984. "Profitable Stream Positions for Salmonids: Relating Specific Growth to Net Energy Gain." *Canadian Journal of Fisheries and Aquatic Sciences* 62:441–51.

Fausch, K. D., C. V. Baxter, and M. Murakami. 2010. "Multiple Stressors in North Temperate Streams: Lessons from Linked Forest-Stream Ecosystems in Northern Japan." *Freshwater Biology* 55:120–34.

Fausch, K. D., C. E. Torgersen, C. V. Baxter, and H. W. Li. 2002. "Landscapes to Riverscapes: Bridging the Gap between Research and Conservation of Stream Fishes." *BioScience* 52:483–98.

Fellows, C. S., H. M. Valett, and C. N. Dahm. 2001. "Whole-Stream Metabolism in Two Montane Streams: Contribution of the Hyporheic Zone." *Limnology and Oceanography* 46:523–31.

Fernández, D., K. Voss, M. Bundschuh, J. P. Zubrod, and R. B. Schäfer. 2015. "Effects of Fungicides on Decomposer Communities and Litter Decomposition in Vineyard Streams." *Science of the Total Environment* 533:40–48.

Ferreira, V., B. Castagneyrol, J. Koricheva, V. Gulis, E. Chauvet, and M. A. S. Graça. 2015. "A Meta-analysis of the Effects of Nutrient Enrichment on Litter Decomposition in Streams." *Biological Reviews* 90:669–88.

Ferreira, V., V. Gulis, and M. A. S. Graça. 2006. "Whole-Stream Nitrate Addition Affects Litter Decomposition and Associated Fungi but Not Invertebrates." *Oecologia* 149:718–29.

Findlay, S. 1995. "Importance of Surface-Subsurface Exchange in Stream Ecosystems: The Hyporheic Zone." *Limnology and Oceanography* 40:159–64.

Finlay, J. C. 2011. "Stream Size and Human Influences on Ecosystem Production in River Networks." *Ecosphere* 2:art87.

Finlay, J. C., J. M. Hood, M. P. Limm, M. E. Power, J. D. Schade, and J. R. Welter. 2011. "Light-Mediated Thresholds in Stream-Water Nutrient Composition in a River Network." *Ecology* 92:140–50.

Finn, D. S., M. T. Bogan, and D. A. Lytle. 2009. "Demographic Stability Metrics for Conservation Prioritization of Isolated Populations." *Conservation Biology* 23:1185–94.

Finn, D. S., N. Bonada, C. Múrria, and J. M. Hughes. 2011. "Small but Mighty: Headwaters Are Vital to Stream Network Biodiversity at Two Levels of Organization." *Journal of the North American Benthological Society* 30:963–80.

Fisher, S. G. 1997. "Creativity, Idea Generation, and the Functional Morphology of Streams." *Journal of the North American Benthological Society* 16:305–18.

Fisher, S. G., L. Gray, N. B. Grimm, and D. E. Busch. 1982. "Temporal Succession in a Desert Stream Ecosystem following Flash Flooding." *Ecological Monographs* 52:93–110.

Fisher, S. G., N. B. Grimm, E. Martí, R. M. Holmes, and J. B. Jones Jr. 1998. "Material Spiraling in Stream Corridors: A Telescoping Ecosystem Model." *Ecosystems* 1:19–34.

Fisher, S. G., and G. E. Likens. 1972. "Stream Ecosystem: Organic Energy Budget." *BioScience* 22:33–35.

Fisher, S. G., and G. E. Likens. 1973. "Energy Flow in Bear Brook, New Hampshire: An Integrative Approach to Stream Ecosystem Metabolism." *Ecological Monographs* 43:421–39.

Fisher, S. G., R. A. Sponseller, and J. B. Heffernan. 2004. "Horizons in Stream Biogeochemistry: Flowpaths to Progress." *Ecology* 85:2369–79.

Flecker, A. S. 1992. "Fish Predation and the Evolution of Invertebrate Drift Periodicity: Evidence from Neotropical Streams." *Ecology* 73:438–48.

Flecker, A. S., Q. Shir, R. M. Almeida, H. Angarita, J. M. Gomes-Selman, R. García-Villacorta, S. A. Sethi et al. 2022. "Reducing Adverse Impacts of Amazon Hydropower Expansion." *Science* 375:753–60.

Flecker, A. S., and C. R. Townsend. 1994. "Community-Wide Consequences of Trout Introduction in New Zealand Streams." *Ecological Applications* 4:798–807.

Flitcroft, R. L., W. R. Brignon, B. Staab, J. R. Bellmore, J. Burnett, P. Burns, B. Cluer et al. 2022. "Rehabilitating Valley Floors to a Stage 0 Condition: A Synthesis of Opening Outcomes." *Frontiers in Environmental Science* 10:892268.

Foley, M. M., F. J. Magilligan, C. E. Torgersen, J. J. Major, C. W. Anderson, P. J. Connolly, D. Wieferich et al. 2017. "Landscape Context and the Biophysical Response of Rivers to Dam Removal in the United States." *PlosOne* 12:e0180107.

Follstad Shah, J. J., J. S. Kominoski, M. Ardón, W. K. Dodds, M. O. Gessner, N. A. Griffiths, C. P. Hawkins et al. 2017. "Global Synthesis of the Temperature Sensitivity of Leaf Litter Breakdown in Streams and Rivers." *Global Change Biology* 23:3064–75.

Forbes, S. A. 1903. "The Food of Fishes." *Illinois Natural History Survey Bulletin* 1:19–70.

Forman, R. T. T., and M. Godron. 1981. "Patches and Structural Components for a Landscape Ecology." *BioScience* 31:733–40.

Forman, R. T. T., and M. Godron. 1986. *Landscape Ecology*. New York: John Wiley and Sons.

Fournier, R. J., G. de Mendoza, R. Sarremejane, and A. Ruhi. 2023. "Isolation Controls Reestablishment Mechanisms and Post-drying Community Structure in an Intermittent Stream." *Ecology* 104:e3911.

Frissell, C. A., W. J. Liss, C. E. Warren, and M. D. Hurley. 1986. "A Hierarchical Framework for Stream Habitat Classification: Viewing Streams in a Watershed Context." *Environmental Management* 10:199–214.

Gardner, J. R., X. Yang, S. N. Topp, M. R. V. Ross, E. H. Altenau, and T. M. Pavelsky. 2021. "The Color of Rivers." *Geophysical Research Letters* 48:e2020GL088946.

Gende, S. M., R. T. Edwards, M. F. Willson, and M. S. Wipfli. 2002. "Pacific Salmon in Aquatic and Terrestrial Ecosystems." *BioScience* 52:917–28.

Genzoli, L., and R. O. Hall Jr. 2016. "Shifts in Klamath River Metabolism following a Reservoir Cyanobacterial Bloom." *Freshwater Science* 35:795–809.

Gessner, M. O., M. A. Bauchrowitz, and M. Escautier. 1991. "Extraction and Quantification of Ergosterol as a Measure of Fungal Biomass in Leaf Litter." *Microbial Ecology* 22:285–91.

Gessner, M. O., and E. Chauvet. 1994. "Importance of Stream Microfungi in Controlling Breakdown Rates of Leaf Litter." *Ecology* 75:1807–17.

Gessner, M. O., and E. Chauvet. 2002. "A Case for Using Litter Breakdown to Assess Functional Stream Integrity." *Ecological Applications* 12:498–510.

Gessner, M. O., E. Chauvet, and M. Dobson. 1999. "A Perspective on Leaf Litter Breakdown in Streams." *Oikos* 85:377–84.

Gessner, M. O., V. Gulis, K. A. Kuehn, E. Chauvet, and K. Suberkropp. 2007. "Fungal Decomposers of Plant Litter in Aquatic Ecosystems." In *The Mycota*, vol. 4, *Environmental and Microbial Relationships*, 2nd ed., edited by C. P. Kubicek and I. S. Druzhinina, 301–24. Berlin: Springer Verlag.

Gessner, M. O., C. M. Swan, C. K. Dang, B. G. McKie, R. D. Bardgett, D. H. Wall, and S. Hättenschwiler. 2010. "Diversity Meets Decomposition." *Trends in Ecology and Evolution* 25:372–80.

Gómez-Gener, L., A. R. Siebers, M. I. Arce, S. Arnon, S. Bernal, R. Bolpagni, T. Datry et al. 2021. "Towards an Improved Understanding of Biogeochemical Processes across Surface-Groundwater Interactions in Intermittent Rivers and Ephemeral Streams." *Earth-Science Reviews* 220:103724.

Gomi, T., R. C. Sidle, R. D. Woodsmith, and M. D. Bryant. 2003. "Characteristics of Channel Steps and Reach Morphology in Headwater Streams, Southeast Alaska." *Geomorphology* 51:225–42.

González, J. M., and M. A. S. Graça. 2003. "Conversion of Leaf Litter to Secondary Production by a Shredding Caddis-Fly." *Freshwater Biology* 48:1578–92.

Gorman, O. T., and J. R. Karr. 1978. "Habitat Structure and Stream Fish Communities." *Ecology* 59:507–15.

Goulding, M. 1980. *The Fishes and the Forest: Explorations in Amazonian Natural History*. Berkeley: University of California Press.

Gounand, I., C. J. Little, E. Harvey, and F. Altermatt. 2018. "Cross-ecosystem Carbon Flows Connecting Ecosystems Worldwide." *Nature Communications* 9:4825.

Graça, M. A. S. 2001. "The Role of Invertebrates on Leaf Litter Decomposition in Streams—A Review." *International Review of Hydrobiology* 86:383–93.

Graça, M. A. S., V. Ferreira, C. Canhoto, A. C. Encalada, F. Guerrero-Bolaño, K. M. Wantzen, and L. Boyero. 2015. "A Conceptual Model of Litter Breakdown in Low Order Streams." *International Review of Hydrobiology* 100:1–12.

Grant, S. B., M. Azizian, P. Cook, F. Boano, and M. A. Rippy. 2018. "Factoring Stream Turbulence into Global Assessments of Nitrogen Pollution." *Science* 359:1266–69.

Gray, L. J., and S. G. Fisher. 1981. "Postflood Recolonization Pathways of Macroinvertebrates in a Lowland Sonoran Desert Stream." *American Midland Naturalist* 106:249–57.

Gregory, S. V. 1978. "Phosphorus Dynamics on Organic and Inorganic Substrates in Streams." *Internationalen Vereinigung für Theoretische und Angewandte Limnologie: Verhandlungen* 20:1340–46.

Gregory, S. V., F. J. Swanson, W. A. McKee, and K. W. Cummins.1991. "An Ecosystem Perspective of Riparian Zones." *BioScience* 41:540–51.

Grill, G., B. Lehner, M. Thieme, B. Geenen, D. Tickner, F. Antonelli, S. Babu et al. 2019. "Mapping the World's Free-Flowing Rivers." *Nature* 569:215–21.

Grimm, N. B. 1987. "Nitrogen Dynamics during Succession in a Desert Stream." *Ecology* 68:1157–70.

Grimm, N. B. 1988. "Role of Macroinvertebrates in Nitrogen Dynamics of a Desert Stream." *Ecology* 69:1884–93.

Grimm, N. B., and S. G. Fisher. 1989. "Stability of Periphyton and Macroinvertebrates to Disturbance by Flash Floods in a Desert Stream." *Journal of the North American Benthological Society* 8:293–307.

Grimm, N. B., and K. C. Petrone. 1997. "Nitrogen Fixation in a Desert Stream Ecosystem." *Biogeochemistry* 37:33–61.

Grubaugh, J. W., and J. B. Wallace. 1995. "Functional Structure and Production of the Benthic Community in a Piedmont River: 1956–1957 and 1991–1992." *Limnology and Oceanography* 40:490–501.

Gulis, V., R. Su, and K. A. Kuehn. 2019. "Fungal Decomposers in Freshwater Environments." In *The Structure and Function of Aquatic Microbial Communities*, edited by C. J. Hurst, 121–55. Advances in Environmental Microbiology 7. Cham, Switzerland: Springer International Publishing.

Gulis, V., and K. Suberkropp. 2003a. "Leaf Litter Decomposition and Microbial Activity in Nutrient-Enriched and Unaltered Reaches of a Headwater Stream." *Freshwater Biology* 48:123–34.

Gulis, V., and K. Suberkropp. 2003b. "Interactions between Stream Fungi and Bacteria Associated with Decomposing Leaf Litter at Different Levels of Nutrient Availability." *Aquatic Microbial Ecology* 30:149–57.

Guo, F., M. J. Kainz, F. Sheldon, and S. E. Bunn. 2016. "The Importance of High-Quality Algal Food Sources in Stream Food Webs—Current Status and Future Perspectives." *Freshwater Biology* 61:815–31.

Gurnell, A. M., G. E. Petts, D. M. Hannah, B. P. Smith, P. J. Edwards, J. Kollmann, J. V. Ward, and K. Tockner. 2001. "Riparian Vegetation and Island Formation along the Gravel-Bed Fiume Tagliamento, Italy." *Earth Surface Processes and Landforms* 26:31–62.

Häder, D. P., and P. W. Barnes. 2019. "Comparing the Impacts of Climate Change on the Responses and Linkages between Terrestrial and Aquatic Ecosystems." *Science of the Total Environment* 682:239–46.

Haggerty, R., E. Martí, A. Argerich, D. von Schiller, and N. B. Grimm. 2009. "Resazurin as a "Smart" Tracer for Quantifying Metabolically Active Transient Storage in Stream Ecosystems." *Journal of Geophysical Research: Biogeosciences* 114: G03014.

Hairston, N. G., F. E. Smith, and L. B. Slobodkin. 1960. "Community Structure, Population Control, and Competition." *American Naturalist* 94:421–25.

Hall, C. A. S. 1972. "Migration and Metabolism in a Temperate Stream Ecosystem." *Ecology* 53:585–604.

Hall, R. O. Jr. 2003. "A Stream's Role in Watershed Nutrient Export." *Proceedings of the National Academy of Sciences* 100:10137–38.

Hall, R. O. Jr., M. A. Baker, E. J. Rosi-Marshall, J. L. Tank, and J. D. Newbold. 2013. "Solute-Specific Scaling of Inorganic Nitrogen and Phosphorus Uptake in Streams." *Biogeosciences* 10:7323–31.

Hall, R. O. Jr., E. S. Bernhardt, and G. E. Likens 2002. "Relating Nutrient Uptake with Transient Storage in Forested Mountain Streams." *Limnology and Oceanography* 47:255–65.

Hall, R. O. Jr., G. E. Likens, and H. M. Malcom. 2001. "Trophic Basis of Invertebrate Production in Two Streams at the Hubbard Brook Experimental Forest." *Journal of the North American Benthological Society* 20:432–47.

Hall, R. O. Jr., and J. L. Tank. 2003. "Ecosystem Metabolism Controls Nitrogen Uptake in Streams in Grand Teton National Park, Wyoming." *Limnology and Oceanography* 48:1120–28.

Hall, R. O. Jr., J. L. Tank, and M. F. Dybdahl. 2003. "Exotic Snails Dominate Nitrogen and Carbon Cycling in a Highly Productive Stream." *Frontiers in Ecology and the Environment* 1:407–11.

Hall, R. O. Jr., J. B. Wallace, and S. L. Eggert. 2000. "Organic Matter Flow in Stream Food Webs with Reduced Detrital Resource Base." *Ecology* 81:3445–63.

Hall, R. O. Jr., C. B. Yackulic, T. A. Kennedy, M. D. Yard, E. J. Rosi-Marshall, N. Voichick, and K. E. Behn. 2015. "Turbidity, Light, Temperature, and Hydropeaking Control Primary Productivity in the Colorado River, Grand Canyon." *Limnology and Oceanography* 60:512–26.

Hansen, L. E., C. B. Yackulic, B. G. Dickson, B. R. Deemer, and R. J. Best. 2023. "Linking Ecosystem Processes to Consumer Growth Rates: Gross Primary Productivity as a Driver of Freshwater Fish Somatic Growth in a Resource-Limited River." *Canadian Journal of Fisheries and Aquatic Sciences* 80:1456–69.

Harper, M. P., and B. L. Peckarsky. 2006. "Emergence Cues of a Mayfly in a High Altitude Stream Ecosystem: Implications for Consequences of Climate Change." *Ecological Applications* 16:612–21.

Harrel, R. C., and T. C. Dorris. 1968. "Stream Order, Morphometry, Physicochemical Conditions, and Community Structure of Benthic Macroinvertebrates in Intermittent Stream System." *American Midland Naturalist* 80:220–51.

Hart, D. D., and C. M. Finelli. 1999. "Physical-Biological Coupling in Streams: The Pervasive Effects of Flow on Benthic Organisms." *Annual Review of Ecology and Systematics* 30:363–95.

Harvey, J. W., and K. E. Bencala. 1993. "The Effect of Streambed Topography on Surface-Subsurface Water Exchange in Mountain Catchments." *Water Resources Research* 29:89–98.

Harvey, J. W., and M. Gooseff. 2015. "River Corridor Science: Hydrologic Exchange and Ecological Consequences from Bedforms to Basins." *Water Resources Research* 51:6893–922.

Hasler, A. D. 1947. "Eutrophication of Lakes by Domestic Drainage." *Ecology* 28:383–95.

Hauer, F. R., H. Locke, V. J. Dreitz, M. Hebblewhite, W. H. Lowe, C. C. Muhlfeld, C. R. Nelson, M. F. Proctor, and S. B. Rood. 2016. "Gravel-Bed River Floodplains Are the Ecological Nexus of Glaciated Mountain Landscapes." *Science Advances* 2:e1600026.

Häussler, J., R. Ryser, and U. Brose. 2021. "Invasive Spread in Meta-Food-Webs Depends on Landscape Structure, Fertilization and Species Characteristics." *Oikos* 130:1257–71.

Hawley, R. J. 2018. "Making Stream Restoration More Sustainable: A Geomorphically, Ecologically, and Socioeconomically Principled Approach to Bridge the Practice with the Science." *BioScience* 68:517–28.

Hayes, D. S., J. M. Brändle, C. Seliger, B. Zeiringer, T. Ferreira, and S. Schmutz. 2018. "Advancing Towards Functional Environmental Flows for Temperate Floodplain Rivers." *Science of the Total Environment* 633:1089–104.

Heffernan, J. B., and M. J. Cohen. 2010. "Direct and Indirect Coupling of Primary Production and Diel Nitrate Dynamics in a Subtropical Spring-Fed River." *Limnology and Oceanography* 55:677–88.

Helfield, J. M., and R. J. Naiman. 2001. "Effects of Salmon-Derived Nitrogen on Riparian Forest Growth and Implications for Stream Productivity." *Ecology* 82:2403–9.

Helton, A. M., G. C. Poole, J. L. Meyer, W. M. Wollheim, B. J. Peterson, P. J. Mulholland, E. S. Bernhardt et al. 2011. "Thinking Outside the Channel: Modeling Nitrogen Cycling in Networked River Ecosystems." *Frontiers in Ecology and the Environment* 9:229–38.

Henry, J. C., and S. G. Fisher, S. G. 2003. "Spatial Segregation of Periphyton Communities in a Desert Stream: Causes and Consequences for N Cycling." *Journal of the North American Benthological Society* 22:511–27.

Hieber, M., and M. O. Gessner. 2002. "Contribution of Stream Detrivores, Fungi, and Bacteria to Leaf Breakdown based on Biomass Estimates." *Ecology* 83:1026–38.

Hildrew, A. G. 2009. "Sustained Research on Stream Communities: A Model System and the Comparative Approach." *Advances in Ecological Research* 41:175–312.

Hildrew, A. G. 2018. *Freshwater Acidification: Natural History, Ecology and Environmental Policy.* Excellence in Ecology 27. Oldendorf, Germany: International Ecology Institute.

Hildrew, A. G., and J. M. Edington. 1979. "Factors Facilitating the Coexistence of Hydropsychid Caddis Larvae (Trichoptera) in the Same River System." *Journal of Animal Ecology* 1:557–76.

Hildrew, A. G., and P. S. Giller. 2023. *The Biology and Ecology of Rivers and Streams*. Oxford, UK: Oxford University Press.

Hildrew, A. G., and C. R. Townsend. 1976. "The Distribution of Two Predators and Their Prey in an Iron-Rich Stream." *Journal of Animal Ecology* 45:41–57.

Hildrew, A. G., and C. R. Townsend. 2007. "Freshwater Biology—Looking Back, Looking Forward." *Freshwater Biology* 52:1863–67.

Hildrew, A. G., C. R. Townsend, and A. Hasham. 1985. "The Predatory Chironomidae of an Iron-Rich Stream: Feeding Ecology and Food Web Structure." *Ecological Entomology* 10:403–13.

Hildrew, A. G., G. Woodward, J. H. Winterbottom, and S. Orton. 2004. "Strong Density Dependence in a Predatory Insect: Large-Scale Experiments in a Stream." *Journal of Animal Ecology* 73:448–58.

Hill, A. R. 1979. "Denitrification in the Nitrogen Budget of a River Ecosystem." *Nature* 281:291–292.

Hoellein, T. J., D. A. Brusewitz, and D. C. Richardson. 2013. "Revisiting Odum (1956): A Synthesis of Aquatic Ecosystem Metabolism." *Limnology and Oceanography* 58:2089–100.

Hogsden, K. L., and J. S. Harding. 2012. "Anthropogenic and Natural Sources of Acidity and Metals and Their Influence on the Structure of Stream Food Webs." *Environmental Pollution* 162:466–74.

Holmes, R. M., S. G. Fisher, and N. B. Grimm. 1994. "Parafluvial Nitrogen Dynamics in a Desert Stream Ecosystem." *Journal of the North American Benthological Society* 13:468–78.

Holmes, R. T., and G. E. Likens 2016. *Hubbard Brook: The Story of a Forested Ecosystem*. New Haven, CT: Yale University Press.

Holt, C. S., and T. F. Waters. 1967. "Effect of Light Intensity on the Drift of Stream Invertebrates." *Ecology* 48:225–34.

Holtgrieve, G. W., D. E. Schindler, T. A. Branch, and Z. T. A'mar. 2010. "Simultaneous Quantification of Aquatic Ecosystem Metabolism and Reaeration Using a Bayesian Statistical Model of Oxygen Dynamics." *Limnology and Oceanography* 55:1047–63.

Hornberger, G. M., and M. G. Kelly. 1975. "Atmospheric Reaeration in a River Using Productivity Analysis." *Journal of the Environmental Engineering Division* 101:729–39.

Hotchkiss, E. R., R. O. Hall Jr., R. A. Sponseller, D. Butman, J. Klaminder, H. Laudon, M. Rosvall, and J. Karlsson. 2015. "Sources of and Processes Controlling $CO_2$ Emissions Change with the Size of Streams and Rivers." *Nature Geoscience* 8:696–99.

Huet, M. 1959. "Profiles and Biology of Western European Streams as Related to Fish Management." *Transactions of the American Fisheries Society* 88:155–63.

Hughes, J. M., P. B. Mather, A. L. Sheldon, and F. W. Allendorf. 1999. "Genetic Structure of the Stonefly, *Yoraperla brevis*, Populations: The Extent of Gene Flow among Adjacent Montane Streams." *Freshwater Biology* 41:63–72.

Hughes, J. M., D. J. Schmidt, and D. S. Finn. 2009. "Genes in Streams: Using DNA to Understand the Movement of Freshwater Fauna and Their Riverine Habitat." *BioScience* 59:573–83.

Humboldt, A. von, and A. Bonpland. 1807. *Essay on the Geography of Plants*. Chicago: University of Chicago Press.

Huryn, A. D. 1996. "An Appraisal of the Allen Paradox in a New Zealand Trout Stream." *Limnology and Oceanography* 41:243–52.

Huryn, A. D. 1998. "Ecosystem-Level Evidence for Top-Down and Bottom-Up Control of Production in a Grassland Stream System." *Oecologia* 115:173–83.

Huryn, A. D., and J. P. Benstead. 2019. "Seasonal Changes in Light Availability Modify the Temperature Dependence of Secondary Production in an Arctic Stream." *Ecology* 100:e02690.

Huryn, A. D., and J. B. Wallace. 1985. "Life History and Production of *Goerita semata* Ross (Trichoptera: Limnephilidae) in the Southern Appalachian Mountains." *Canadian Journal of Zoology* 63:2604–11.

Huryn, A. D., and J. B. Wallace. 2019. "Habitat, Life History, Secondary Production, and Behavioral Adaptations of Aquatic Insects." In *An Introduction to the Aquatic Insects of North America*, 5th ed., edited by R. W. Merritt, K. W. Cummins, and M. B. Berg, 65–116. Dubuque, IA: Kendall Hunt Publishing.

Huston, M. 1979. "A General Hypothesis of Species Diversity." *American Naturalist* 113:81–101.

Hutchinson, G. E. 1967. *A Treatise on Limnology*. New York: John Wiley and Sons.

Hutchinson, G. E., and V. T. Bowen. 1947. "A Direct Demonstration of the Phosphorus Cycle in a Small Lake." *Proceedings of the National Academy of Sciences* 33:148–53.

Hynes, H. B. N. 1960. *The Biology of Polluted Waters*. Liverpool: Liverpool University Press.

Hynes, H. B. N. 1961. "The Invertebrate Fauna of a Welsh Mountain Stream." *Archiv für Hydrobiologie* 57:344–88.

Hynes, H. B. N. 1963. "Imported Organic Matter and Secondary Productivity in Streams." *Proceedings of the 16th International Congress of Zoology* 16:324–29.

Hynes, H. B. N. 1970. *The Ecology of Running Waters*. Toronto, Canada: University of Toronto Press.

Hynes, H. B. N. 1975. "The Stream and Its Valley." *Internationale Vereinigung für theoretische und angewandte Limnologie: Verhandlungen* 19:1–15.

Hynes, H. B. N. 1981. "Hynes, H. B. N. 1961. 'The Invertebrate Fauna of a Welsh Mountain Stream.' *Archiv für Hydrobiologie* 57:344–88." *Current Contents* 32:164.

Hynes, H. B. N. 2001. *Nunc Dimittis: A Life in the River of Time*. Kerkwerve, Netherlands: Backhuys Publishers.

Hynes, H. B. N., and T. R. Williams. 1962. "The Effect of DDT on the Fauna of a Central African Stream." *Annals of Tropical Medicine and Parasitology* 56:78–91.

Ide, F. P. 1942. "Availability of Aquatic Insects as Food for the Speckled Trout, *Salvelinus fontinalis*." *Transactions of the North American Wildlife Conference* 7:442–50.

Illies, J., and L. Botosaneanu. 1963. "Problèmes et méthodes de la classification et de la zonation écologique des eaux courantes, considerées surtout du point de vue faunistique." *Internationalen Vereinigung für Theoretische und Angewandte Limnologie: Mitteilungen* 12:1–57.

Irons, J. G. III, M. W. Oswood, R. J. Stout, and C. M. Pringle. 1994. "Latitudinal Patterns in Leaf Litter Breakdown: Is Temperature Really Important?" *Freshwater Biology* 32:401–11.

Jabiol, J., A. Lecerf, S. Lamothe, M. O. Gessner, and E. Chauvet. 2019. "Litter Quality Modulates Effects of Dissolved Nitrogen on Leaf Decomposition by Stream Microbial Communities." *Microbial Ecology* 77:959–66.

Jackson, J. K., and S. G. Fisher. 1986. "Secondary Production, Emergence, and Export of Aquatic Insects of a Sonoran Desert Stream." *Ecology* 67:629–38.

Jardine, T. D., A. W. Galloway, and M. J. Kainz. 2020. "Unlocking the Power of Fatty Acids as Dietary Tracers and Metabolic Signals in Fishes and Aquatic Invertebrates." *Philosophical Transactions of the Royal Society B* 375:20190639.

Jellyman, D. J., G. J. Glova, M. L. Bonnett, A. I. McKerchar, and K. R. Allen. 2000. *The Horokiwi Stream 50 Years On: A Study of the Loss of a Productive Trout Fishery*. NIWA Technical Report 83. Wellington, New Zealand: National Institute of Water and Atmospheric Research.

Johnson, B. R., and J. B. Wallace. 2005. "Bottom-Up Limitation of a Stream Salamander in a Detritus-Based Food Web." *Canadian Journal of Fisheries and Aquatic Sciences* 62:301–11.

Johnson, M. F., C. R. Thorne, J. M. Castro, G. M. Kondolf, C. S. Mazzacano, S. B. Rood, and C. Westbrook. 2020. "Biomic River Restoration: A New Focus for River Management." *River Research and Applications* 36:3–12.

Johnson, N. M., G. E. Likens, F. H. Bormann, D. W. Fisher, and R. S. Pierce. 1969. "A Working Model for the Variation in Stream Water Chemistry at the Hubbard Brook Experimental Forest, New Hampshire." *Water Resources Research* 5:1353–63.

Jones, J. B., S. G. Fisher, and N. B. Grimm. 1995a. "Nitrification in the Hyporheic Zone of a Desert Stream Ecosystem." *Journal of the North American Benthological Society* 14:249–58.

Jones, J. B., S. G. Fisher, and N. B. Grimm. 1995b. "Vertical Hydrologic Exchange and Ecosystem Metabolism in a Sonoran Desert Stream." *Ecology* 76:942–52.

Jones, J. B., and P. J. Mulholland. 2000. *Streams and Groundwaters*. New York: Academic Press.

Jones, J. R. E. 1949. "A Further Ecological Study of Calcareous Streams in the 'Black Mountain' District of South Wales." *Journal of Animal Ecology* 18:142–59.

Jones, J. R. 1950. "A Further Ecological Study of the River Rheidol: The Food of the Common Insects of the Main-Stream." *Journal of Animal Ecology* 19:159–74.

Jones, J. R. E. 1951. "An Ecological Study of the River Towy." *Journal of Animal Ecology* 20:68–86.

Joober, R., N. Schmitz, L. Annable, and P. Boksa. 2012. "Publication Bias: What Are the Challenges and Can They Be Overcome?" *Journal of Psychiatry and Neuroscience* 37:149–52.

Jordan, M. I., and T. M. Mitchell. 2015. "Machine Learning: Trends, Perspectives, and Prospects." *Science* 349:255–60.

Juday, C., J. M. Blair, and E. F. Wilda. 1943. "The Photosynthetic Activities of the Aquatic Plants of Little John Lake, Vilas County, Wisconsin." *American Midland Naturalist* 30:426–46.

Junk, W. J., P. B. Bayley, and R. E. Sparks 1989. "The Flood Pulse Concept in River-Floodplain Systems." *Canadian Journal of Fisheries and Aquatic Sciences* Special Publication 106:110–27.

Junker, J. R., W. F. Cross, J. P. Benstead, A. D. Huryn, J. M. Hood, D. Nelson, G. M. Gíslason, and J. S. Ólafsson. 2020. "Resource Supply Governs the Apparent Temperature Dependence of Animal Production in Stream Ecosystems." *Ecology Letters* 23:1809–19.

Kale, V. S., L. L. Ely, Y. Enzel, and V. R. Baker. 1994. "Geomorphic and Hydrologic Aspects of Monsoon Floods on the Narmada and Tapi Rivers in Central India." In *Geomorphology and Natural Hazards*, edited by M. Morisawa, 157–68. Amsterdam: Elsevier Science.

Kareiva, P., M. Marvier, and M. McClure. 2000. "Recovery and Management Options for Spring/Summer Chinook Salmon in the Columbia River Basin." *Science* 290:977–79.

Karr, J. R. 1991. "Biological Integrity: A Long-Neglected Aspect of Water Resource Management." *Ecological Applications* 1:66–84.

Karrenberg, S., P. J. Edwards, and J. Kollmann. 2002. "The Life History of Salicaceae Living in the Active Zone of Floodplains." *Freshwater Biology* 47:733–48.

Kasahara, T., and S. M. Wondzell. 2003. "Geomorphic Controls on Hyporheic Exchange Flow in Mountain Streams." *Water Resources Research* 39:1005.

Kaushik, N. K. 1982. "Kaushik, N. K., and H. B. N. Hynes. 1971. 'The Fate of the Dead Leaves That Fall into Streams.' *Archiv für Hydrobiologie* 68:465–515." *Current Contents* 33:20.

Kaushik, N. K., and H. B. N. Hynes. 1968. "Experimental Study on the Role of Autumn Shed Leaves in Aquatic Environments." *Journal of Ecology* 56:229–43.

Kaushik, N. K., and H. B. N. Hynes. 1971. "The Fate of the Dead Leaves That Fall into Streams." *Archiv für Hydrobiologie* 68:465–515.

Kelleher, C., A. Ward, J. L. A. Knapp, P. J. Blaen, M. J. Kurz, J. D. Drummond, J. P. Zarnetske et al. 2019. "Exploring Tracer Information and Model Framework Trade-Offs to Improve Estimation of Stream Transient Storage Processes." *Water Resources Research* 55:3481–501.

Keller, E. A., and F. J. Swanson. 1979. "Effects of Large Organic Material on Channel Form and Fluvial Processes." *Earth Surface Processes* 4:361–80.

Kemp, M. J., and W. K. Dodds. 2001. "Centimeter-Scale Patterns in Dissolved Oxygen and Nitrification Rates in a Prairie Stream." *Journal of the North American Benthological Society* 20:347–57.

Kennedy, T. A., J. D. Muehlbauer, C. B. Yackulic, D. A. Lytle, S. W. Miller, K. L. Dibble, E. W. Kortenhoeven, A. N. Metcalfe, and C. V. Baxter. 2016. "Flow Management for Hydropower Extirpates Aquatic Insects, Undermining River Food Webs." *BioScience* 66:561–75.

King, D. L., and R. C. Ball. 1967. "Comparative Energetics of a Polluted Stream." *Limnology and Oceanography* 12:27–33.

Koenig, L. E., A. M. Helton, P. Savoy, E. Bertuzzo, J. B. Heffernan, R. O. Hall Jr., and E. S. Bernhardt. 2019. "Emergent Productivity Regimes of River Networks." *Limnology and Oceanography* 4:173–81.

Kohler, S. L. 1983. "Positioning on Substrates, Positioning Changes, and Diel Drift Periodicities in Mayflies." *Canadian Journal of Zoology* 61:1362–68.

Kohler, S. L., and M. J. Wiley. 1997. "Pathogen Outbreaks Reveal Large-Scale Effects of Competition in Stream Communities." *Ecology* 78:2164–76.

Kolkwitz, R., and M. Marsson. 1902. "Grundsätze für die biologische Beurteilung des Wassers nach seiner Flora und Fauna." *Mitteilungen aus der Königlichen Prüfungsanstalt für Wasserversorgung und Abwasserbeseitigung zu Berlin* 1:33–72.

Kominoski, J. S., J. J. Follstad Shah, C. Canhoto, D. G. Fischer, D. P. Giling, E. González, N. A. Griffiths et al. 2013. "Forecasting Functional Implications of Global Changes in Riparian Plant Communities." *Frontiers in Ecology and the Environment* 11:423–32.

Kondolf, G. M., M. W. Smeltzer, and S. F. Railsback. 2001. "Design and Performance of a Channel Reconstruction Project in a Coastal California Gravel-Bed Stream." *Environmental Management* 28:761–76.

Krabbenhoft, C. A., G. H. Allen, P. Lin, S. E. Godsey, D. C. Allen, R. M. Burrows, A. G. DelVecchia et al. 2022. "Assessing Placement Bias of the Global River Gauge Network." *Nature Sustainability* 5:586–92.

Kuehne, R. A. 1962. "A Classification of Streams, Illustrated by Fish Distribution in an Eastern Kentucky Creek." *Ecology* 43:608–14.

Kupferberg, S. J., J. C. Marks, and M. E. Power. 1994. "Effects of Variation in Natural Algal and Detrital Diets on Larval Anuran (*Hyla regilla*) Life-History Traits." *Copeia* 1994:446–57.

Lake, P. S. 2000. "Disturbance, Patchiness, and Diversity in Streams." *Journal of the North American Benthological Society* 19:573–92.

Lake, P. S., N. Bond, and P. Reich. 2007. "Linking Ecological Theory with Stream Restoration." *Freshwater Biology* 52:597–615.

Lamberti, G. A., and V. H. Resh. 1983. "Stream Periphyton and Insect Herbivores: An Experimental Study of Grazing by a Caddisfly Population." *Ecology* 64:1124–35.

Lancaster, J., and B. J. Downes. 2018. "Aquatic versus Terrestrial Insects: Real or Presumed Differences in Population Dynamics?" *Insects* 9:157.

Lancaster, J., B. J. Downes, and A. Arnold. 2010. "Environmental Constraints on Oviposition Limit Egg Supply of a Stream Insect at Multiple Scales." *Oecologia* 163:373–84.

Lancaster, J., and A. G. Hildrew. 1993. "Flow Refugia and the Microdistribution of Lotic Macroinvertebrates." *Journal of the North American Benthological Society* 12:385–93.

Langbein, W. B. and L. B. Leopold. 1964. "Quasi-equilibrium States in Channel Morphology." *American Journal of Science* 262:782–94.

Larsen, S., J. D. Muehlbauer, and E. Martí. 2016. "Resource Subsidies between Stream and Terrestrial Ecosystems under Global Change." *Global Change Biology* 22:2489–504.

Lawton, J. H. 1999. "Are There General Laws in Ecology?" *Oikos* 84:177–92.

Layer, K., A. G. Hildrew, and G. Woodward. 2013. "Grazing and Detritivory in 20 Stream Food Webs across a Broad pH Gradient." *Oecologia* 171:459–71.

Lecerf, A. 2017. "Methods for Estimating the Effect of Litterbag Mesh Size on Decomposition." *Ecological Modelling* 362:65–68.

Ledger, M. E., L. E. Brown, F. K. Edwards, A. M. Milner, and G. Woodward. 2013. "Drought Alters the Structure and Functioning of Complex Food Webs." *Nature Climate Change* 3:223–27.

Leigh, C., A. J. Boulton, J. L. Courtwright, K. Fritz, C. L. May, R. H. Walker, and T. Datry. 2016. "Ecological Research and Management of Intermittent Rivers: An Historical Review and Future Directions." *Freshwater Biology* 61:1181–99.

Leopold, A. 1949. *A Sand County Almanac*. Oxford, UK: Oxford University Press.

Leopold, L. B., M. G. Wolman, and J. P. Miller. 1964. *Fluvial Processes in Geomorphology*. San Francisco, CA: W. H. Freeman.

Likens, G. E. 1984. "Beyond the Shoreline: A Watershed-Ecosystem Approach." *SIL Proceedings, 1922–2010* 22:1–22.

Likens, G. E., and F. H. Bormann. 1974. "Linkages between Terrestrial and Aquatic Ecosystems." *BioScience* 24:447–56.

Likens, G. E., F. H. Bormann, N. M. Johnson, D. W. Fisher, and R. S. Pierce. 1970. "Effects of Forest Cutting and Herbicide Treatment on Nutrient Budgets in the Hubbard Brook Watershed-Ecosystem." *Ecological Monographs* 40:23–47.

Likens, G. E., F. H. Bormann, R. S. Pierce, J. S. Eaton, and N. M. Johnson. 1977. *Biogeochemistry of a Forested Catchment*. New York: Springer,

Likens, G. E., F. H. Bormann, R. S. Pierce, and W. A. Reiners. 1978. "Recovery of a Deforested Ecosystem." *Science* 199:492–96.

Lindeman, R. L. 1942. "The Trophic-Dynamic Aspect of Ecology." *Ecology* 23:399–417.

Lindenmayer, D. B., and G. E. Likens. 2018. "Maintaining the Culture of Ecology." *Frontiers in Ecology and the Environment* 16:195.

Lindenmayer, D. B., G. E. Likens, C. J. Krebs, and R. J. Hobbs. 2010. "Improved Probability of Detection of Ecological 'Surprises.'" *Proceedings of the National Academy of Sciences* 107:21957–62.

Linduska, J. P. 1942. "Bottom Type as a Factor Influencing the Local Distribution of Mayfly Nymphs." *Canadian Entomologist* 74:26–30.

LINX Collaborators. 2014. "The Lotic Intersite Nitrogen Experiments: An Example of Successful Ecological Research Collaboration." *Freshwater Science* 33:700–710.

Livers, B., and E. Wohl. 2015. "An Evaluation of Stream Characteristics in Glacial versus Fluvial Process Domains in the Colorado Front Range." *Geomorphology* 231:72–82.

Lovett, G. M., J. J. Cole, and M. L. Pace. 2006. "Is Net Ecosystem Production Equal to Ecosystem Carbon Accumulation?" *Ecosystems* 9:152–55.

Lowe, R. L. 2015. "Ruth Patrick: The River Doctor." *Limnology and Oceanography Bulletin* 24:108–11.

Lowe, W. H., and G. E. Likens. 2005. "Moving Headwater Streams to the Head of the Class." *BioScience* 55:196–97.

Lowrance, R., R. Todd, J. Fail Jr., O. Hendrickson Jr., R. Leonard, and L. Asmussen. 1984. "Riparian Forests as Nutrient Filters in Agricultural Watersheds." *BioScience* 34:374–77.

Lytle, D. A. 2002. "Flash Floods and Aquatic Insect Life-History Evolution: Evaluation of Multiple Models." *Ecology* 83:370–85.

Lytle, D. A., and N. L. Poff. 2004. "Adaptation to Natural Flow Regimes." *Trends in Ecology and Evolution* 19:94–100.

Macan, T. T. 1958. "Methods of Sampling the Bottom Fauna in Stony Streams." *Internationale Vereinigung für Theoretische und Angewandte Limnologie: Mitteilungen* 8:1–21.

Macan, T. T. 1962. "Biotic Factors in Running Waters." *Schweizerische Zeitschrift für Hydrologie* 24:386–407.

Macneale, K. H., B. L. Peckarsky, and G. E. Likens. 2005. "Stable Isotopes Identify Dispersal Patterns of Stonefly Populations Living along Stream Corridors." *Freshwater Biology* 50:1117–30.

Malard, F., K. Tockner, M. J. Dole-Olivier, and J. V. Ward. 2002. "A Landscape Perspective of Surface-Subsurface Hydrological Exchanges in River Corridors." *Freshwater Biology* 47:621–40.

Maloney, K. O., K. P. Krause, C. Buchanan, L. E. Hay, G. J. McCabe, Z. M. Smith, T. L. Sohl, and J. A. Young. 2020. "Disentangling the Potential Effects of Land-Use and Climate Change on Stream Conditions." *Global Change Biology* 26:2251–69.

Mann, K. H. 1965. "Energy Transformations by a Population of Fish in the River Thames." *Journal of Animal Ecology* 34:253–75.

Manning, D. W. P., A. L. Dere, A. W. Miller, and T. J. Coleman. 2022. "Evidence for Pulse-Shunt Carbon Exports from a Mixed Land-Use, Restored Prairie Watershed." *Freshwater Science* 41:284–98.

Marcarelli, A. M., C. V. Baxter, J. R. Benjamin, Y. Miyake, M. Murakami, K. D. Fausch, and S. Nakano. 2020. "Magnitude and Direction of Stream Forest Community Interactions Change with Time Scale." *Ecology* 101:e03064.

Marcarelli, A. M., C. V. Baxter, M. M. Mineau, and R. O. Hall Jr. 2011. "Quantity and Quality: Unifying Food Web and Ecosystem Perspectives on the Role of Resource Subsidies in Freshwaters." *Ecology* 92:1215–25.

Marczak, L. B., and J. S. Richardson. 2007. "Spiders and Subsidies: Results from the Riparian Zone of a Coastal Temperate Rainforest." *Journal of Animal Ecology* 76:687–94.

Margalef, R. 1960. "Ideas for a Synthetic Approach to the Ecology of Running Waters." *Internationale Revue der gesamten Hydrobiologie und Hydrographie* 45:133–53.

Margalef, R. 1963. "On Certain Unifying Principles in Ecology." *American Naturalist* 97:357–74.

Marks, J. C. 2019. "Revisiting the Fates of Dead Leaves That Fall into Streams." *Annual Review of Ecology, Evolution, and Systematics* 50:547–68.

Marzolf, E. R., P. J. Mulholland, and A. D. Steinman. 1994. "Improvements to the Diurnal Upstream-Downstream Dissolved Oxygen Change Technique for Determining Whole-Stream Metabolism in Small Streams." *Canadian Journal of Fisheries and Aquatic Sciences* 51:1591–99.

Mathews, C. P., and A. Kowalczewski. 1969. "The Disappearance of Leaf Litter and Its Contribution to Production in the River Thames." *Journal of Ecology* 57:543–52.

May, R. M. 1973. *Stability and Complexity in Model Ecosystems*. Princeton, NJ: Princeton University Press.

Mayer, M. S., and G. E. Likens. 1987. "The Importance of Algae in a Shaded Headwater Stream as Food for an Abundant Caddisfly (Trichoptera)." *Journal of the North American Benthological Society* 6:262–69.

McAllister, J. W. 1996. *Beauty and Revolution in Science*. New York: Cornell University Press.

McAuliffe, J. R. 1984. "Competition for Space, Disturbance, and the Structure of a Benthic Stream Community." *Ecology* 65:894–908.

McClain, M. E., E. W. Boyer, C. L. Dent, S. E. Gergel, N. B. Grimm, P. M. Groffman, S. C. Hart et al. 2003. "Biogeochemical Hot Spots and Hot Moments at the Interface of Terrestrial and Aquatic Ecosystems." *Ecosystems* 6:301–12.

McColl, R. H. S. 1974. "Self-Purification of Small Freshwater Streams: Phosphate, Nitrate, and Ammonia Removal." *New Zealand Journal of Marine and Freshwater Research* 8:375–88.

McCutchan, J. H., and W. M. Lewis. 2002. "Relative Importance of Carbon Sources for Macroinvertebrates in a Rocky Mountain Stream." *Limnology and Oceanography* 47:742–52.

McDowell, D. M., and R. J. Naiman. 1986. "Structure and Function of a Benthic Invertebrate Stream Community as Influenced by Beaver (*Castor canadensis*)." *Oecologia* 68:481–89.

McIntire, C. D., and H. K. Phinney. 1965. "Laboratory Studies of Periphyton Production and Community Metabolism in Lotic Environments." *Ecological Monographs* 35:237–58.

McIntosh, A. R., H. S. Greig, and S. Howard. 2022. "Regulation of Open Populations of a Stream Insect through Larval Density Dependence." *Journal of Animal Ecology* 91:1582–95.

McIntosh, A. R., and C. R. Townsend. 1994. "Interpopulation Variation in Mayfly Antipredator Tactics: Differential Effects of Contrasting Predatory Fish." *Ecology* 75:2078–90.

McIntosh, A. R., and C. R. Townsend. 1995. "Impacts of an Introduced Predatory Fish on Mayfly Grazing in New Zealand Streams." *Limnology and Oceanography* 40:1508–12.

McIntosh, A. R., and C. R. Townsend. 1996. "Interactions between Fish, Grazing Invertebrates and Algae in a New Zealand Stream: A Trophic Cascade Mediated by Fish-Induced Changes to Grazer Behaviour?" *Oecologia* 108:174–81.

McPeek, M. A., and B. L. Peckarsky. 1998. "Life Histories and the Strengths of Species Interactions: Combining Mortality, Growth and Fecundity Effects." *Ecology* 79:235–47.

Mecom, J. O., and K. W. Cummins. 1964. "A Preliminary Study of the Trophic Relationships of the Larvae of *Brachycentrus americanus* (Banks) (Trichoptera: Brachycentridae)." *Transactions of the American Microscopical Society* 83:233–43.

Meffe, G. K. 1984. "Effects of Abiotic Disturbance on Coexistence of Predator-Prey Fish Species." *Ecology* 65:1525–34.

Meffe, G. K., and W. L. Minckley. 1987. "Persistence and Stability of Fish and Invertebrate Assemblages in a Repeatedly Disturbed Sonoran Desert Stream." *American Midland Naturalist* 117:177–91.

Meitzen, K. M., M. W. Doyle, M. C. Thoms, and C. E. Burns. 2013. "Geomorphology within the Interdisciplinary Science of Environmental Flows." *Geomorphology* 200:143–54.

Melillo, J. M., J. D. Aber, and J. F. Muratore. 1982. "Nitrogen and Lignin Control of Hardwood Leaf Litter Decomposition Dynamics." *Ecology* 63:621–26.

Merritt, D. M., M. L. Scott, N. L. Poff, G. T. Auble, and D. A. Lytle. 2010. "Theory, Methods and Tools for Determining Environmental Flows for Riparian Vegetation: Riparian Vegetation-Flow Response Guilds." *Freshwater Biology* 55:206–25.

Meyer, J. L. 1989. "Can P/R Ratio Be Used to Assess the Food Base of Stream Ecosystems? A Comment on Rosenfeld and Mackay." *Oikos* 54:119–121.

Meyer, J. L. 2013. "Streams and Their Valleys." In *Fundamentals of Ecosystem Science*, edited by K. C. Weathers, K. L. Strayer, and G. E. Likens, 267–73. San Francisco, CA: Academic Press, Elsevier.

Meyer, J. L., A. C. Benke, R. T. Edwards, and J. B. Wallace. 1997. "Organic Matter Dynamics in the Ogeechee River, a Blackwater River in Georgia, USA." *Journal of the North American Benthological Society* 16:82–87.

Meyer, J. L., and R. T. Edwards. 1990. "Ecosystem Metabolism and Turnover of Organic Carbon along a Blackwater River Continuum." *Ecology* 71:668–77.

Meyer, J. L., and G. E. Likens. 1979. "Transport and Transformation of Phosphorus in a Forest Stream Ecosystem." *Ecology* 60:1255–69.

Meyer, J. L., D. L. Strayer, J. B. Wallace, S. L. Eggert, G. S. Helfman, and N. E. Leonard. 2007. "The Contribution of Headwater Streams to Biodiversity in River Networks." *Journal of the American Water Resources Association* 43:86–103.

Meyer, J. L., and J. B. Wallace. 2001. "Lost Linkages and Lotic Ecology: Rediscovering Small Streams." In *Ecology: Achievement and Challenge*, edited by M. C. Press, N. J. Huntley, and S. Levin, 295–317. The 41st Symposium of the British Ecological Society. Orlando, FL: Blackwell Science.

Milan, D. J. 2012. "Geomorphic Impact and System Recovery following an Extreme Flood in an Upland Stream: Thinhope Burn, Northern England, UK." *Geomorphology* 138:319–28.

Minckley, W. L. 1963. "The Ecology of a Spring Stream, Doe Run, Meade County, Kentucky." *Wildlife Monographs* 11:1–124.

Minshall, G. W. 1967. "Role of Allochthonous Detritus in the Trophic Structure of a Woodland Springbrook Community." *Ecology* 48:139–49.

Minshall, G. W. 1978. "Autotrophy in Stream Ecosystems." *BioScience* 28:767–71.

Minshall, G. W., R. C. Petersen, K. W. Cummins, T. L. Bott, J. R. Sedell, C. E. Cushing, and R. L. Vannote. 1983. "Interbiome Comparison of Stream Ecosystem Dynamics." *Ecological Monographs* 53:1–25.

Molles, M. C., C. S. Crawford, L. M. Ellis, H. M. Valett, and C. N. Dahm. 1998. "Managed Flooding for Riparian Ecosystem Restoration." *BioScience* 48:749–56.

Montgomery, D. R. 1999. "Process Domains and the River Continuum." *Journal of the American Water Resources Association* 35:397–410.

Montgomery, D. R., and J. M. Buffington. 1997. "Channel-Reach Morphology in Mountain Drainage Basins." *Geological Society of America Bulletin* 109:596–611.

Moulton, T. P., C. M. Andrade, and V. Neres-Lima. 2019. "The Outcome of an Exclusion Experiment Depends on the Method: Shrimps, Shredders and Leaf Breakdown in a Tropical Stream." *Freshwater Science* 38:131–41.

Mouron, S., D. Eme, A. Bellec, M. Bertrand, S. Mammola, F. Liébault, C. J. Douady, and F. Malard. 2022. "Unique and Shared Effects of Local and Catchment Predictors over Distribution of Hyporheic Organisms: Does the Valley Rule the Stream?" *Ecography* 2022:e06099.

Mulholland, P. J. 1992. "Regulation of Nutrient Concentrations in a Temperate Forest Stream: Roles of Upland, Riparian, and Instream Processes." *Limnology and Oceanography* 37:1512–26.

Mulholland, P. J., A. M. Helton, G. C. Poole, R. O. Hall Jr., S. K. Hamilton, B. J. Peterson, J. L. Tank et al. 2008. "Stream Denitrification across Biomes and Its Response to Anthropogenic Nitrate Loading." *Nature* 452:202–5.

Mulholland, P. J., J. D. Newbold, J. W. Elwood, L. A. Ferren, and J. R. Webster. 1985. "Phosphorus Spiraling in a Woodland Stream: Seasonal Variations." *Ecology* 66:1012–23.

Mulholland, P. J., J. L. Tank, D. M. Sanzone, W. M. Wollheim, B. J. Peterson, J. R. Webster, and J. L. Meyer. 2000. "Nitrogen Cycling in a Forest Stream Determined by a $^{15}N$ Tracer Addition." *Ecological Monographs* 70:471–93.

Müller, K. 1954. "Investigations on the Organic Drift in North Swedish Streams." *Report of the Institute of Freshwater Research, Drottningholm* 35:133–48.

Müller, K. 1974. "Stream Drift as a Chronological Phenomenon in Running Water Ecosystems." *Annual Review of Ecology and Systematics* 5:309–23.

Müller, K. 1982. "The Colonization Cycle of Freshwater Insects." *Oecologia* 53:202–7.

Murray, A. B., M. A. F. Knaapen, M. Tal, and M. L. Kirwan. 2008. "Biomorphodynamics: Physical-Biological Feedbacks That Shape Landscapes." *Water Resources Research* 44.

Naegeli, M. W., and U. Uehlinger. 1997. "Contribution of the Hyporheic Zone to Ecosystem Metabolism in a Prealpine Gravel-Bed-River." *Journal of the North American Benthological Society* 16:794–804.

Naiman, R. J. 1983. "The Annual Pattern and Spatial Distribution of Aquatic Oxygen Metabolism in Boreal Forest Watersheds." *Ecological Monographs* 53:73–94.

Naiman, R. J., H. Décamps, and M. E. McClain. 2010. *Riparia: Ecology, Conservation, and Management of Streamside Communities*. Academic Press, Elsevier.

Naiman, R. J., J. M. Melillo, and J. E. Hobbie. 1986. "Ecosystem Alteration of Boreal Forest Streams by Beaver (*Castor canadensis*)." *Ecology* 67:1254–69.

Nakamura, K., K. Tockner, and K. Amano. 2006. "River and Wetland Restoration: Lessons from Japan." *BioScience* 56:419–29.

Nakano, S., H. Miyasaka, and N. Kuhara. 1999. "Terrestrial-Aquatic Linkages: Riparian Arthropod Inputs Alter Trophic Cascades in a Stream Food Web." *Ecology* 80:2435–41.

Nakano, S., and M. Murakami. 2001. "Reciprocal Subsidies: Dynamic Interdependence between Terrestrial and Aquatic Food Webs." *Proceedings of the National Academy of Sciences* 98:166–70.

Naman, S. M., J. S. Rosenfeld, and J. S. Richardson. 2016. "Causes and Consequences of Invertebrate Drift in Running Waters: From Individuals to Populations and Trophic Fluxes." *Canadian Journal of Fisheries and Aquatic Sciences* 73:1292–305.

Needham, P. R. 1928. "A Net for the Capture of Stream Drift Organisms." *Ecology* 9:339–42.

Needham, P. R. 1938. *Trout Streams.* Ithaca, NY: Comstock Publishing Company.

Needham, P. R., and R. L. Usinger. 1956. "Variability in the Macrofauna of a Single Riffle in Prosser Creek, California, as Indicated by the Surber Samples." *Hilgardia* 24:383–409.

Nelson, D., J. P. Benstead, A. D. Huryn, W. F. Cross, J. M. Hood, P. W. Johnson, J. R. Junker, G. M. Gíslason, and J. S. Ólafsson. 2020. "Thermal Niche Diversity and Trophic Redundancy Drive Neutral Effects of Warming on Energy Flux through a Stream Food Web." *Ecology* 101:e02952.

Nelson, D. J., and. D. C. Scott. 1962. "Role of Detritus in the Productivity of a Rock-Outcrop Community in a Piedmont Stream." *Limnology and Oceanography* 7:396–413.

Newbold, J. D., J. W. Elwood, R. V. O'Neill, and A. L. Sheldon. 1983. "Phosphorus Dynamics in a Woodland Stream Ecosystem: A Study of Nutrient Spiraling." *Ecology* 64:1249–65.

Newbold, J. D., J. W. Elwood, R. V. O'Neill, and W. Van Winkle. 1981. "Measuring Nutrient Spiralling in Streams." *Canadian Journal of Fisheries and Aquatic Sciences* 38:860–63.

Newbold, J. D., J. W. Elwood, M. S. Schulze, R. W. Stark, and J. C. Barmeier. 1983. "Continuous Ammonium Enrichment of a Woodland Stream—Uptake Kinetics, Leaf Decomposition, and Nitrification." *Freshwater Biology* 13:193–204.

Newbold, J. D., P. J. Mulholland, J. W. Elwood, and R. V. O'Neill. 1982. "Organic Carbon Spiralling in Stream Ecosystems." *Oikos* 38:266–72.

Newell, S. Y., T. L. Arsuffi, and R. D. Fallon. 1988. "Fundamental Procedures for Determining Ergosterol Content of Decaying Plant-Material by Liquid-Chromatography." *Applied and Environmental Microbiology* 54:1876–79.

Nielsen, J. M., E. L. Clare, B. Hayden, M. T. Brett, and P. Kratina. 2018. "Diet Tracing in Ecology: Method Comparison and Selection." *Methods in Ecology and Evolution* 9:278–91.

Nielsen, L. P., P. B. Christensen, N. P. Revsbech, and J. Sørensen. 1990. "Denitrification and Photosynthesis in Stream Sediment Studied with Microsensor and Wholecore Techniques." *Limnology and Oceanography* 35:1135–44.

Nilsson, C., R. L. Brown, R. Jansson, and D. M. Merritt. 2010. "The Role of Hydrochory in Structuring Riparian and Wetland Vegetation." *Biological Reviews* 85:837–58.

Norris, K. S. 1998. *Mountain Time: Reflections on the Wild World and Our Place in It.* Oakland, CA: University of California Natural Reserve System.

Norris, R. H., and C. P. Hawkins. 2000. "Monitoring River Health." *Hydrobiologia* 425:5–17.

Noss, R. F., and A. Y. Cooperrider. 1994. *Saving Nature's Legacy: Protecting and Restoring Biodiversity.* Washington, DC: Island Press.

Nuttle, W. K. 2002. "Is Ecohydrology One Idea or Many?" *Hydrological Sciences Journal* 47:805–7.

Oberg, K., and D. S. Mueller. 2007. "Validation of Streamflow Measurements Made with Acoustic Doppler Current Profilers." *Journal of Hydraulic Engineering* 133:1421–32.

Odum, E. P. 1969. "The Strategy of Ecosystem Development." *Science* 164:262–70.

Odum, H. T. 1956. "Primary Production in Flowing Waters." *Limnology and Oceanography* 1:102–17.

Odum, H. T. 1957. "Trophic Structure and Productivity of Silver Springs, Florida." *Ecological Monographs* 27:55–112.

Odum, H. T., and E. P. Odum. 1955. "Trophic Structure and Productivity of a Windward Coral Reef Community on Eniwetok Atoll." *Ecological Monographs* 25:291–320.

O'Gorman, E. J., O. L. Petchey, K. J. Faulkner, B. Gallo, T. A. Gordon, J. Neto-Cerejeira, J. S. Ólafsson, D. E. Pichler, M. S. Thompson, and G. Woodward. 2019. "A Simple Model Predicts How Warming Simplifies Wild Food Webs." *Nature Climate Change* 9:611–16.

Oksanen, L., S. D. Fretwell, J. Arruda, and P. Niemela. 1981. "Exploitation Ecosystems in Gradients of Primary Productivity." *American Naturalist* 118:240–61.

Orghidan, T. 1955. "Un nou domeniu de viata acvatica subterana 'Biotopul hiporeic.'" *Buletin Stiintific sectia de Biologie si stiinte Agricole si sectia de Geologie si Geografie* 7:657–76.

Orghidan, T. 2010. "A New Habitat of Subsurface Waters: The Hyporheic Biotope." *Fundamental and Applied Limnology* 176:291–302.

Pace, M. L., S. R. Carpenter, J. J. Cole, J. J. Coloso, J. F. Kitchell, J. R. Hodgson, J. J. Middelburg et al. 2007. "Does Terrestrial Organic Carbon Subsidize the Planktonic Food Web in a Clear-Water Lake?" *Limnology and Oceanography* 52:2177–89.

Packman, A. I., and M. Salehin. 2003. "Relative Roles of Stream Flow and Sedimentary Conditions in Controlling Hyporheic Exchange." *Hydrobiologia* 494:291–97.

Paetzold, A., J. F. Bernet, and K. Tockner. 2006. "Consumer-Specific Responses to Riverine Subsidy Pulses in a Riparian Arthropod Assemblage." *Freshwater Biology* 51:1103–15.

Paine, R. T. 1966. "Foodweb Complexity and Species Diversity." *American Naturalist* 100:65–75.

Paine, R. T. 1969. "A Note on Trophic Complexity and Community Stability." *American Naturalist* 103:91–93.

Palmer, M. A., K. L. Hondula, and B. J. Koch. 2014. "Ecological Restoration of Streams and Rivers: Shifting Strategies and Shifting Goals." *Annual Review of Ecology, Evolution, and Systematics* 45:247–69.

Palmer, M. A., C. A. R. Liermann, C. Nilsson, M. Flörke, J. Alcamo, P. S. Lake, and N. Bond. 2008. "Climate Change and the World's River Basins: Anticipating Management Options." *Frontiers in Ecology and the Environment* 6:81–89.

Palmer, M. A., and A. Ruhi. 2019. "Linkages between Flow Regime, Biota, and Ecosystem Processes: Implications for River Restoration." *Science* 365:eaaw2087.

Paola, C., E. Foufoula-Georgiou, W. E. Dietrich, M. Hondzo, D. Mohrig, G. Parker, M. E. Power, I. Rodriguez-Iturbe, V. Voller, and P. Wilcock. 2006. "Toward a Unified Science of the Earth's Surface: Opportunities for Synthesis among Hydrology, Geomorphology, Geochemistry, and Ecology." *Water Resources Research* 42.

Parker, S. M., and A. D. Huryn. 2006. "Food Web Structure and Function in Two Arctic Streams with Contrasting Disturbance Regimes." *Freshwater Biology* 51:1249–63.

Pascoal, C., and F. Cássio. 2004. "Contribution of Fungi and Bacteria to Leaf Litter Decomposition in a Polluted River." *Applied and Environmental Microbiology* 70:5266–73.

Patrick, R. 1949. "A Proposed Biological Measure of Stream Conditions, Based on a Survey of the Conestoga Basin, Lancaster County, Pennsylvania." *Proceedings of the Academy of Natural Sciences of Philadelphia* 101:277–341.

Peckarsky, B. L. 1980. "Predator-Prey Interactions between Stoneflies and Mayflies: Behavioral Observations." *Ecology* 61:932–43.

Peckarsky, B. L. 1983. "Biotic Interactions or Abiotic Limitations? A Model of Lotic Community Structure." In *Dynamics of Lotic Ecosystems*, edited by T. D. Fontaine and S. M. Bartell, 303–23. Ann Arbor, MI: Ann Arbor Science.

Peckarsky, B. L., J. D. Allan, A. R. McIntosh, and B. W. Taylor. 2010. "Understanding the Role of Predation in Open Systems: The Value of Place-Based Research." In *The Ecology of Place*, edited by I. A. Billick and M. V. Price, 185–206. Chicago: University of Chicago Press.

Peckarsky, B. L., A. C. Encalada, and A. R. McIntosh. 2011. "Why Do Vulnerable Mayflies Thrive in Trout Streams?" *American Entomologist* 57:152–64.

Peckarsky, B. L., B. L. Kerans, A. R. McIntosh, and B. W. Taylor. 2008. "Predator Effects on Prey Population Dynamics in Open Systems." *Oecologia* 156:431–40.

Peckarsky, B. L., A. R. McIntosh, B. W. Taylor, and J. Dahl. 2002. "Predator Chemicals Induce Changes in Mayfly Life History Traits: A Whole-Stream Manipulation." *Ecology* 83:612–18.

Peckarsky, B. L., and M. A. Penton. 1989. "Mechanisms of Prey Selection by Stream-Dwelling Stoneflies." *Ecology* 70:1203–18.

Peckarsky, B. L., B. W. Taylor, and C. C. Caudill. 2000. "Hydrologic and Behavioral Constraints on Oviposition of Stream Insects: Implications for Adult Dispersal." *Oecologia* 125:186–200.

Peckarsky, B. L., B. W. Taylor, A. R. McIntosh, M. A. McPeek, and D. A. Lytle. 2001. "Variation in Mayfly Size at Metamorphosis as a Developmental Response to Risk of Predation." *Ecology* 82:740–57.

Pennak, R. W., and E. D. Van Gerpen. 1947. "Bottom Fauna Production and Physical Nature of the Substrate in a Northern Colorado Trout Stream." *Ecology* 28:42–48.

Peterjohn, W. T., and D. L. Correll. 1984. "Nutrient Dynamics in an Agricultural Watershed: Observations on the Role of a Riparian Forest." *Ecology* 65:1466–75.

Petersen, R. C., and K. W. Cummins. 1974. "Leaf Processing in a Woodland Stream." *Freshwater Biology* 4:343–68.

Petersen, R. C., and K. W. Cummins. 1991. "Petersen, R. C., and K. W. Cummins. 1974. 'Leaf Processing in a Woodland Stream.' *Freshwater Biology* 4:343–68." *Current Contents* 28:10.

Peterson, B. J., M. Bahr, and G. W. Kling. 1997. "A Tracer Investigation of Nitrogen Cycling in a Pristine Tundra River." *Canadian Journal of Fisheries and Aquatic Sciences* 54:2361–67.

Peterson, B. J., J. E. Hobbie, A. E. Hershey, M. A. Lock, T. E. Ford, J. R. Vestal, V. L. McKinley et al. 1985. "Transformation of a Tundra River from Heterotrophy to Autotrophy by Addition of Phosphorus." *Science* 229:1383–86.

Peterson, B. J., W. M. Wollheim, P. J. Mulholland, J. R. Webster, J. L. Meyer, J. L. Tank, E. Martí et al. 2001. "Control of Nitrogen Export from Watersheds by Headwater Streams." *Science* 292:86–90.

Petts, G. E. 1984. *Impounded Rivers: Perspectives for Ecological Management*. Chichester: Wiley.

Petts, G. E., and A. M. Gurnell. 2005. "Dams and Geomorphology: Research Progress and Future Directions." *Geomorphology* 71:27–47.

Pickett, S. T. A., and P. S. White. 1985. *The Ecology of Natural Disturbance and Patch Dynamics*. New York: Academic Press.

Pimm, S. L., and R. L. Kitching. 1987. "The Determinants of Food Chain Length." *Oikos* 50:302–7.

Pinay, G., V. J. Black, A. M. Planty-Tabacchi, B. Gumiero, and H. Décamps. 2000. "Geomorphic Control of Denitrification in Large River Floodplain Soils." *Biogeochemistry* 50:163–82.

Pinay, G., L. Roques, and A. Fabre. 1993. "Spatial and Temporal Patterns of Denitrification in a Riparian Forest." *Journal of Applied Ecology* 581–91.

Plont, S., J. Riney, and E. R. Hotchkiss. 2022. "Integrating Perspectives on Dissolved Organic Carbon Removal and Whole-Stream Metabolism." *Journal of Geophysical Research: Biogeosciences* 127:E2021JG006610.

Poff, N. L. 1992. "Why Disturbances Can Be Predictable: A Perspective on the Definition of Disturbance in Streams." *Journal of the North American Benthological Society* 11:86–92.

Poff, N. L. 1997. "Landscape Filters and Species Traits: Towards Mechanistic Understanding and Prediction in Stream Ecology." *Journal of the North American Benthological Society* 16:391–409.

Poff, N. L., J. D. Allan, M. B. Bain, J. R. Karr, K. L. Prestegaard, B. D. Richter, R. E. Sparks, and J. C. Stromberg. 1997. "The Natural Flow Regime: A Paradigm for River Conservation and Restoration." *BioScience* 47:769–84.

Poff, N. L., B. P. Bledsoe, and C. O. Cuhaciyan. 2006. "Hydrologic Variation with Land Use across the Contiguous United States: Geomorphic and Ecological Consequences for Stream Ecosystems." *Geomorphology* 79:264–85.

Poff, N. L., and J. H. Matthews. 2013. "Environmental Flows in the Anthropocene: Past Progress and Future Prospects." *Current Opinion in Environmental Sustainability* 5:667–75.

Poff, N. L., J. D. Olden, D. M. Merritt, and D. M. Pepin. 2007. "Homogenization of Regional River Dynamics by Dams and Global Biodiversity Implications." *Proceedings of the National Academy of Sciences* 104:5732–37.

Poff, N. L., J. D. Olden, N. K. M. Vieira, D. S. Finn, M. P. Simmons, and B. C. Kondratieff. 2006. "Functional Trait Niches of North American Lotic Insects: Traits-Based Ecological Applications in Light of Phylogenetic Relationships." *Journal of the North American Benthological Society* 25:730–55.

Poff, N. L., R. E. Tharme, and A. H. Arthington. 2017. "Evolution of Environmental Flows Assessment Science, Principles, and Methodologies." In *Water for the Environment*, edited by A. C. Horne, J. A. Webb, M. J. Stewardson, B. Richter, and M. Acreman, 203–36. Academic Press, Elsevier.

Poff, N. L., and J. V. Ward. 1989. "Implications of Streamflow Variability and Predictability for Lotic Community Structure—A Regional-Analysis of Streamflow Patterns." *Canadian Journal of Fisheries and Aquatic Sciences* 46:1805–18.

Poff, N. L., and J. V. Ward. 1990. "Physical Habitat Template of Lotic Systems: Recovery in the Context of Historical Pattern of Spatiotemporal Heterogeneity." *Environmental Management* 14:629–45.

Polis, G. A. 1991. "Trophic Interactions in Deserts: An Empirical Critique of Food-Web Theory." *American Naturalist* 138:123–55.

Polis, G. A., W. B. Anderson, and R. D. Holt. 1997. "Toward an Integration of Landscape and Food Web Ecology: The Dynamics of Spatially Subsidized Food Webs." *Annual Review of Ecology and Systematics* 28:289–316.

Pollock, M. M., T. J. Beechie, J. M. Wheaton, C. E. Jordan, N. Bouwes, N. Weber, and C. Volk. 2014. "Using Beaver Dams to Restore Incised Stream Ecosystems." *BioScience* 64:279–90.

Polvi, L. E., E. E. Wohl, and D. M. Merritt. 2011. "Geomorphic and Process Domain Controls on Riparian Zones in the Colorado Front Range." *Geomorphology* 125:504–16.

Poole, G. C. 2002. "Fluvial Landscape Ecology: Addressing Uniqueness within the River Discontinuum." *Freshwater Biology* 47:641–60.

Poole, G. C. 2010. "Stream Hydrogeomorphology as a Physical-Science Basis for Advances in Stream Ecology." *Journal of the North American Benthological Society* 29:12–25.

Poole, G. C., S. J. O. Daniel, K. L. Jones, W. W. Woessner, and E. S. Bernhardt. 2008. "Hydrologic Spiraling: The Role of Multiple Interactive Flow Paths in Stream Ecosystems." *River Research and Applications* 1031:1018–31.

Power, M. E. 1990. "Effects of Fish in River Food Webs." *Science* 250:811–14.

Power, M. E., T. L. Dudley, and S. D. Cooper. 1989. "Grazing Catfish, Fishing Birds, and Attached Algae in a Panamanian Stream." *Environmental Biology of Fishes* 26:285–94.

Power, M. E., R. L. Lowe, P. C. Furey, J. Welter, M. Limm, J. C. Finlay, C. Bode, S. Chang, M. Goodrich, and J. Sculley. 2009. "Algal Mats and Insect Emergence in

Rivers under Mediterranean Climates: Towards Photogrammetric Surveillance." *Freshwater Biology* 54:2101–15.

Power, M. E., J. Marks, and M. S. Parker. 1992. "Variation in the Vulnerability of Prey to Different Predators: Community-Level Consequences." *Ecology* 73:2218–23.

Power, M. E., W. J. Matthews, and A. J. Stewart. 1985. "Grazing Minnows, Piscivorous Bass, and Stream Algae: Dynamics of a Strong Interaction." *Ecology* 66:1448–56.

Power, M. E., M. S. Parker, and W. E. Dietrich. 2008. "Seasonal Reassembly of a River Food Web: Floods, Droughts, and Impacts of Fish." *Ecological Monographs* 78:263–82.

Power, M. E., M. S. Parker, and J. T. Wootton. 1996. "Disturbance and Food Chain Length in Rivers." In *Food Webs: Integration of Patterns and Dynamics*, edited by G. A. Polis and K. O. Winemiller, 286–97. New York: Chapman and Hall.

Power, M. E., W. E. Rainey, M. S. Parker, J. Sabo, A. Smyth, S. Khandwala, J. C. Finlay et al. 2004. "River-to-Watershed Subsidies in an Old-Growth Conifer Forest." In *Food Webs at the Landscape Level*, edited by G. A. Polis, M. E. Power, and G. Huxel, 217–40. Chicago: University of Chicago Press.

Power, M. E., A. Sun, G. Parker, W. E. Dietrich, and J. T. Wootton. 1995. "Hydraulic Food-Chain Models: An Approach to the Study of Food-Web Dynamics in Large Rivers." *BioScience* 45:159–67.

Priest, H. J., and D. S. Finn. 2023. "Bucketloads of Aquatic Invertebrates in a Dry Intermittent Stream." *Ecology* 104:e3936.

Pringle, C. M. 1997. "Exploring How Disturbance Is Transmitted Upstream: Going Against the Flow." *Journal of the North American Benthological Society* 16:425–38.

Pringle, C. M., and G. A. Blake. 1994. "Quantitative Effects of Atyid Shrimp (Decapoda: Atyidae) on the Depositional Environment in a Tropical Stream: Use of Electricity for Experimental Exclusion." *Canadian Journal of Fisheries and Aquatic Sciences* 51:1443–50.

Pringle, C. M., R. J. Naiman, G. Bretschko, J. R. Karr, M. W. Oswood, J. R. Webster, R. L. Welcomme, and M. J. Winterbourn. 1988. "Patch Dynamics in Lotic Systems: The Stream as a Mosaic." *Journal of the North American Benthological Society* 7:503–24.

Pringle, R. M., and M. C. Hutchinson. 2020. "Resolving Food-Web Structure." *Annual Review of Ecology and Systematics* 51:55–80.

Raymond, P. A., J. Hartmann, R. Lauerwald, S. Sobek, C. McDonald, M. Hoover, D. Butman et al. 2013. "Global Carbon Dioxide Emissions from Inland Waters." *Nature* 503:355–59.

Raymond, P. A., J. E. Saiers, and W. V. Sobczak. 2016. "Hydrological and Biogeochemical Controls on Watershed Dissolved Organic Matter Transport: Pulse-Shunt Concept." *Ecology* 97:5–16.

Regester, K. J., K. R. Lips, and M. R. Whiles. 2006. "Energy Flow and Subsidies Associated with the Complex Life Cycle of Ambystomatid Salamanders in Ponds and Adjacent Forest in Southern Illinois." *Oecologia* 147:303–14.

Resh, V. H., A. V. Brown, A. P. Covich, M. E. Gurtz, H. W. Li, G. W. Minshall, S. R. Reice, A. L. Sheldon, J. B. Wallace, and R. C. Wissmar. 1988. "The Role of Disturbance in Stream Ecology." *Journal of the North American Benthological Society* 7:433–55.

Resh, V. H., A. G. Hildrew, B. Statzner, and C. R. Townsend. 1994. "Theoretical Habitat Templets, Species Traits, and Species Richness: A Synthesis of Long-Term Ecological Research on the Upper Rhône River in the Context of Concurrently Developed Ecological Theory." *Freshwater Biology* 31:539–54.

Rice, S. P. 2021. "Why So Skeptical? The Role of Animals in Fluvial Geomorphology." *Wiley Interdisciplinary Reviews, Water* 8:e1549.

Rice, S. P., M. T. Greenwood, and C. B. Joyce. 2001. "Tributaries, Sediment Sources, and the Longitudinal Organisation of Macroinvertebrate Fauna along River Systems." *Canadian Journal of Fisheries and Aquatic Sciences* 58:824–40.

Richards, C., R. Haro, L. Johnson, and G. Host. 1997. "Catchment and Reach-Scale Properties as Indicators of Macroinvertebrate Species Traits." *Freshwater Biology* 37:219–30.

Richardson, J. S. 1992. "Food, Microhabitat, or Both? Macroinvertebrate Use of Leaf Accumulations in a Montane Stream." *Freshwater Biology* 27:169–76.

Richardson, R. E. 1921. "Changes in the Bottom and Shore Fauna of the Middle Illinois River and Its Connecting Lakes since 1913 to 1915 as a Result of Increase, Southward, of Sewage Pollution." *Bulletin of the Illinois Natural History Survey* 14:33–75.

Richter, B. D., J. V. Baumgartner, R. Wigington, and D. P. Braun. 1997. "How Much Water Does a River Need?" *Freshwater Biology* 37:231–49.

Riley, G. A. 1940. "Limnological Studies in Connecticut. III. The Plankton of Linsley Pond." *Ecological Monographs* 10:280–306.

Riley, G. A. 1953. "Letter to the Editor." *Journal du Conseil Permanent International pour l'Exploration de la Mer* 19:85–89.

Roberts, B. J., and P. J. Mulholland. 2007. "In-Stream Biotic Control on Nutrient Biogeochemistry in a Forested Stream, West Fork of Walker Branch." *Journal of Geophysical Research: Biogeosciences* 112:G04002.

Rocher-Ros, G., R. A. Sponseller, W. Lidberg, C.-M. Mörth, and R. Giesler. 2019. "Landscape Process Domains Drive Patterns of $CO_2$ Evasion from River Networks." *Limnology and Oceanography Letters* 4:87–95.

Rode, M., S. Halbedel, M. R. Anis, D. Borchardt, and M. Weitere. 2016. "Continuous In-Stream Assimilatory Nitrate Uptake from High-Frequency Sensor Measurements." *Environmental Science and Technology* 50:5685–94.

Rodriguez-Iturbe, I., and A. Rinaldo. 1997. *Fractal River Basins: Chance and Self-Organization*. New York: Cambridge University Press.

Rood, S. B., C. R. Gourley, E. M. Ammon, L. G. Heki, J. R. Klotz, M. L. Morrison, D. Mosley, G. G. Scoppettone, S. Swanson, and P. L. Wagner. 2003. "Flows for Floodplain Forests: A Successful Riparian Restoration." *BioScience* 53:647–56.

Rosemond, A. D., J. P. Benstead, P. M. Bumpers, V. Gulis, J. S. Kominoski, D. W. P. Manning, K. Suberkropp, and J. B. Wallace. 2015. "Experimental Nutrient Additions Accelerate Terrestrial Carbon Loss from Stream Ecosystems." *Science* 347:1142–45.

Rosemond, A. D., P. J. Mulholland, and J. W. Elwood. 1993. "Top-Down and Bottom-Up Control of Stream Periphyton: Effects of Nutrients and Herbivores." *Ecology* 74:1264–80.

Rosemond, A. D., C. M. Pringle, A. Ramirez, and M. J. Paul. 2001. "A Test of Top-Down and Bottom-Up Control in a Detritus-Based Food Web." *Ecology* 82:2279–93.

Rosenfeld, J. S., and R. J. Mackay. 1987. "Assessing the Food Base of Stream Ecosystems: Alternatives to the P/R Ratio." *Oikos* 50:141–47.

Rosi-Marshall, E. J., H. A. Wellard Kelly, R. O. Hall Jr., and K. A. Vallis. 2016. "Methods for Quantifying Aquatic Macroinvertebrate Diets." *Freshwater Science* 35:229–36.

Ross, H. H. 1963. "Stream Communities and Terrestrial Biomes." *Archiv für Hydrobiologie* 59:235–42.

Ross, M. R. V., S. N. Topp, A. P. Appling, X. Yang, C. Kuhn, D. Butman, M. Simard, and T. M. Pavelsky. 2019. "AquaSat: A Data Set to Enable Remote Sensing of Water Quality for Inland Waters." *Water Resources Research* 55:10012–25.

Rounick, J. S., M. J. Winterbourn, and G. L. Lyon. 1982. "Differential Utilization of Allochthonous and Autochthonous Inputs by Aquatic Invertebrates in some New Zealand Streams: A Stable Carbon Isotope Study." *Oikos* 39:191–98.

Rüegg, J., C. C. Conn, E. P. Anderson, T. J. Battin, E. S. Bernhardt, M. Boix Canadell, S. M. Bonjour et al. 2020. "Thinking Like a Consumer: Linking Aquatic Basal Metabolism and Consumer Dynamics." *Limnology and Oceanography* 64:1–17.

Rundio, D. E., and S. T. Lindley. 2012. "Reciprocal Fluxes of Stream and Riparian Invertebrates in a Coastal California Basin with Mediterranean Climate." *Ecological Research* 27:539–50.

Runkel, R. L., and S. C. Chapra. 1993. "An Efficient Numerical Solution of the Transient Storage Equations for Solute Transport in Small Streams." *Water Resources Research* 29:211–15.

Ryther, J. H. 1954. "The Ratio of Photosynthesis to Respiration in Marine Plankton Algae and Its Effect upon the Measurement of Productivity." *Deep Sea Research* 2:134–39.

Sabo, J. L., and M. E. Power. 2002. "River-Watershed Exchange: Effects of Riverine Subsidies on Riparian Lizards and Their Terrestrial Prey." *Ecology* 83:1860–69.

Sagoff, M. 2019. "Will NEON Kill Ecology?" *Issues in Science and Technology* 35:54–62.

Sánchez-Montoya, M. M., T. Datry, A. Ruhi, S. M. Carlson, R. Corti, and K. Tockner. 2023. "Intermittent Rivers and Ephemeral Streams Are Pivotal Corridors for Aquatic and Terrestrial Animals." *BioScience* 73:291–301.

Sargent, M. C., and T. S. Austin. 1949. "Organic Productivity of an Atoll." *Eos: Transactions of the American Geophysical Union* 30:245–49.

Savoy, P., A. P. Appling, J. B. Heffernan, E. G. Stets, J. S. Read, J. W. Harvey, and E. S. Bernhardt. 2019. "Metabolic Rhythms in Flowing Waters: An Approach for Classifying River Productivity Regimes." *Limnology and Oceanography* 64:1835–51.

Schade, J. D., K. MacNeill, S. A. Thomas, F. C. McNeely, J. R. Welter, J. Hood, M. Goodrich, M. E. Power, and J. C. Finlay. 2011. "The Stoichiometry of Nitrogen and Phosphorus Spiralling in Heterotrophic and Autotrophic Streams." *Freshwater Biology* 56:424–36.

Schade, J. D., J. R. Welter, E. Martí, and N. B. Grimm. 2005. "Hydrologic Exchange and N Uptake by Riparian Vegetation in an Arid-Land Stream." *Journal of the North American Benthological Society* 24:19–28.

Schindler, D. W. 1977. "Evolution of Phosphorus Limitation in Lakes: Natural Mechanisms Compensate for Deficiencies of Nitrogen and Carbon in Eutrophied Lakes." *Science* 195:260–62.

Schlosser, I. J. 1982. "Fish Community Structure and Function along Two Habitat Gradients in a Headwater Stream." *Ecological Monographs* 52:395–414.

Schlosser, I. J. 1987. "The Role of Predation in Age- and Size-Related Habitat Use by Stream Fishes." *Ecology* 68:651–59.

Schlosser, I. J. 1991. "Stream Fish Ecology: A Landscape Perspective." *BioScience* 41:704–12.

Schmid-Araya, J. M., A. G. Hildrew, A. Robertson, P. E. Schmid, and J. Winterbottom. 2002. "The Importance of Meiofauna in Food Webs: Evidence from an Acid Stream." *Ecology* 83:1271–85.

Schmid-Araya, J. M., and P. E. Schmid. 2000. "Trophic Relationships: Integrating Meiofauna into a Realistic Benthic Food Web." *Freshwater Biology* 44:149–63.

Schubel, J. R., C. C. Conrad, D. Debinski, P. M. Kareiva, G. I. Matsumoto, D. M. McKnight, C. Parmesan et al. 2014. *Enhancing the Value and Sustainability of Field Stations and Marine Laboratories in the 21st Century.* Washington, DC: National Research Council of the National Academies of Science.

Schumm, S. A. 1977. *The Fluvial System.* New York: John Wiley and Sons.

Scott, D. 1958. "Ecological Studies on the Trichoptera of the River Dean, Cheshire." *Archiv für Hydrobiologie* 54:340–92.

Sculley, J. B., R. L. Lowe, C. A. Nittrouer, T. M. Drexler, and M. E. Power. 2017. "Eighty Years of Food Web Response to Interannual Variation in Discharge Recorded in River Diatom Frustules from an Ocean Sediment Core." *Proceedings of the National Academy of Sciences* 114:10155–59.

Sedell, J. R., and J. L. Froggatt. 1984. "Importance of Streamside Forests to Large Rivers: The Isolation of the Willamette River, Oregon, USA, from Its Floodplain by Snagging and Streamside Forest Removal." *Internationale Vereinigung für theoretische und angewandte Limnologie: Verhandlungen* 22:1828–34.

Seena, S., F. Bärlocher, O. Sobral, M. O. Gessner, D. Dudgeon, B. G. McKie, E. Chauvet et al. 2019. "Biodiversity of Leaf Litter Fungi in Streams along a Latitudinal Gradient." *Science of the Total Environment* 661:306–15.

Segatto, P. L., T. J. Battin, and E. Bertuzzo. 2021. "The Metabolic Regimes at the Scale of an Entire Stream Network Unveiled through Sensor Data and Machine Learning." *Ecosystems* 24:1792–1809.

Sentis, A., J. M. Montoya, and M. Lurgi. 2021. "Warming Indirectly Increases Invasion Success in Food Webs." *Proceedings of the Royal Society B* 288:20202622.

Shama, L. N., K. B. Kubow, J. Jokela, and C. T. Robinson. 2011. "Bottlenecks Drive Temporal and Spatial Genetic Changes in Alpine Caddisfly Metapopulations." *BMC Evolutionary Biology* 11:1–11.

Sheldon, A. L. 1968. "Species Diversity and Longitudinal Succession in Stream Fishes." *Ecology* 49:193–98.

Simon, A., M. Doyle, M. Kondolf, F. D. Shields Jr., B. Rhoads, and M. McPhillips. 2007. "Critical Evaluation of How the Rosgen Classification and Associated 'Natural Channel Design' Methods Fail to Integrate and Quantify Fluvial Processes and Channel Response." *Journal of the American Water Resources Association* 43:1117–31.

Sommerfield, C. K., A. S. Ogston, B. L. Mullenbach, D. E. Drake, C. R. Alexander, C. A. Nittrouer, J. C. Borgeld, R. A. Wheatcroft, and E. L. Leithold. 2007. "Oceanic Dispersal and Accumulation of River Sediment." In *Continental Margin Sedimentation: From Sediment Transport to Sequence Stratigraphy*, edited by Charles A. Nittrouer, James A. Austin, Michael E. Field, Joseph H. Kravitz, James P. M. Syvitski, and Patricia L. Wiberg, 157–212. Malden, MA: Wiley.

Sosnovsky, A., M. E. Lallement, M. Rechencq, E. E. Zattara, M. V. Fernández, S. Leiva, M. J. Suárez, R. Daga, and R. J. C. Cantet. 2022. "The Influence of Topography and Land Use on Hydrological and Nutrient Dynamics in two Andean Streams from Northern Patagonia." *New Zealand Journal of Marine and Freshwater Research* 56:78–97.

Sousa, W. P. 1984. "The Role of Disturbance in Natural Communities." *Annual Review of Ecology and Systematics* 15:353–91.

Southwood, T. R. E. 1977. "Habitat, the Templet for Ecological Strategies? Presidential Address to British Ecological Society, 5 January 1977." *Journal of Animal Ecology* 46:337–65.

Sprules, W. M. 1947. "An Ecological Investigation of Stream Insects in Algonquin Park, Ontario." *University of Toronto Studies Biological Series* 56:1–81.

Stanford, J. A., L. C. Alexander, and D. C. Whited. 2017. "Riverscapes." In *Methods in Stream Ecology*, vol. 1, edited by F. R. Hauer and G. A. Lamberti, 3–20. Cambridge, MA: Academic Press, Elsevier.

Stanford, J. A., M. S. Lorang, and F. R. Hauer. 2005. "The Shifting Habitat Mosaic of River Ecosystems." *Internationale Vereinigung für theoretische und angewandte Limnologie: Verhandlungen* 29:123–36.

Stanford, J. A., and J. V. Ward. 1993. "An Ecosystem Perspective of Alluvial Rivers: Connectivity and the Hyporheic Corridor." *Journal of the North American Benthological Society* 12:48–60.

Stanford, J. A., J. V. Ward, W. J. Liss, C. A. Frissell, R. N. Williams, J. A. Lichatowich, and C. C. Coutant. 1996. "A General Protocol for Restoration of Regulated Rivers." *Regulated Rivers: Research & Management* 12:391–413.

Stanley, E. H., S. G. Fisher, and N. B. Grimm. 1997. "Ecosystem Expansion and Contraction in Streams." *BioScience* 47:427–35.

Stanley, E. H., S. M. Powers, and N. R. Lottig. 2010. "The Evolving Legacy of Disturbance in Stream Ecology: Concepts, Contributions, and Coming Challenges." *Journal of the North American Benthological Society* 29:67–83.

Statzner, B., J. A. Gore, and V. H. Resh. 1988. "Hydraulic Stream Ecology: Observed Patterns and Potential Applications." *Journal of the North American Benthological Society* 7:307–60.

Statzner, B., and B. Higler. 1985. "Questions and Comments on the River Continuum Concept." *Canadian Journal of Fisheries and Aquatic Sciences* 42:1038–44.

Statzner, B., V. H. Resh, and A. L. Roux. 1994. "The Synthesis of Long-Term Ecological Research in the Context of Concurrently Developed Ecological Theory: Design of a Research Strategy for the Upper Rhône River and Its Floodplain." *Freshwater Biology* 31:253–63.

Stelzer, R. S., J. Heffernan, G. E. Likens. 2003. "The Influence of Dissolved Nutrients and Particulate Organic Matter Quality on Microbial Respiration and Biomass in a Forest Stream." *Freshwater Biology* 48:1925–37.

Stewart, I. T., D. R. Cayan, and M. D. Dettinger. 2005. "Changes toward Earlier Streamflow Timing across Western North America." *Journal of Climate* 18:1136–55.

Stockner, J. G., and K. R. S. Shortreed. 1978. "Enhancement of Autotrophic Production by Nutrient Addition in a Coastal Rainforest Stream on Vancouver Island." *Journal of the Fisheries Board of Canada* 35:28–34.

Stouffer. D. B., J. Camacho, W. Jiang, and L. A. Nunes Amaral. 2007. "Evidence for the Existence of a Robust Pattern of Prey Selection in Food Webs." *Proceedings of the Royal Society B* 274:1931–40.

Stout, J. 1980. "Leaf Decomposition Rates in Costa Rican Lowland Tropical Rainforest Streams." *Biotropica* 12:264–72.

Strayer, D. L., and D. Dudgeon. 2010. "Freshwater Biodiversity Conservation: Recent Progress and Future Challenges." *Journal of the North American Benthological Society* 29:344–58.

Stream Solute Workshop. 1990. "Concepts and Methods for Assessing Solute Dynamics in Stream Ecosystems." *Journal of the North American Benthological Society* 9:95–119.

Su, X., L. E. Polvi, L. Lind, F. Pilotto, and C. Nilsson. 2019. "Importance of Landscape Context for Post-restoration Recovery of Riparian Vegetation." *Freshwater Biology* 64:1015–28.

Subalusky, A. L., and D. M. Post. 2019. "Context Dependency of Animal Resource Subsidies." *Biological Reviews* 94:517–38.

Suberkropp, K., and E. Chauvet. 1995. "Regulation of Leaf Breakdown by Fungi in Streams: Influences of Water Chemistry." *Ecology* 76:1433–45.

Suberkropp, K., and M. J. Klug. 1976. "Fungi and Bacteria Associated with Leaves during Processing in a Woodland Stream." *Ecology* 57:707–19.

Swan, C. M., L. Boyero, and C. Canhoto, eds. 2021. *The Ecology of Plant Litter Decomposition in Stream Ecosystems.* Cham, Switzerland: Springer-Nature Switzerland.

Tanentzap, A. J., B. W. Kielstra, G. M. Wilkinson, M. Berggren, N. Craig, P. A. Del Giorgio, J. Grey et al. 2017. "Terrestrial Support of Lake Food Webs: Synthesis Reveals Controls over Cross-Ecosystem Resource Use." *Science Advances* 3:e1601765.

Tank, J. L., J. L. Meyer, D. M. Sanzone, P. J. Mulholland, J. R. Webster, B. J. Peterson, W. M. Wollheim, and N. E. Leonard. 2000. "Analysis of Nitrogen Cycling in a Forest Stream during Autumn Using a ¹⁵N-Tracer Addition." *Limnology and Oceanography* 45:1013–29.

Tank, J. L., E. J. Rosi-Marshall, M. A. Baker, and R. O. Hall Jr. 2008. "Are Rivers Just Big Streams? A Pulse Method to quantify nitrogen demand in a large river." *Ecology* 89:2935–45.

Tank, J. L., E. J. Rosi-Marshall, N. A. Griffiths, S. A. Entrekin, M. L. Stephen. 2010. "A Review of Allochthonous Organic Matter Dynamics and Metabolism in Streams." *Journal of the North American Benthological Society* 29:118–46.

Tank, J. L., and J. R. Webster. 1998. "Interaction of Substrate and Nutrient Availability on Wood Biofilm Processes in Streams." *Ecology* 79:2168–79.

Tank, J. L., J. R. Webster, E. F. Benfield, and R. L. Sinsabaugh. 1998. "Effect of Leaf Litter Exclusion on Microbial Enzyme Activity Associated with Wood Biofilms in Streams." *Journal of the North American Benthological Society* 17:95–103.

Tavares-Cromer, A. F., and. D. D. Williams. 1996. "The Importance of Temporal Resolution in Food Web Analysis: Evidence from Detritus-Based Stream." *Ecological Monographs* 66:91–113.

Teal, J. M. 1957. "Community Metabolism in a Temperate Cold Spring." *Ecological Monographs* 27:283–302.

Thienemann, A. 1912. "Der Bergbach des Sauerlandes. Faunistisch-biologische Untersuchungen." *Internationale Revue der gesamten Hydrobiologie und Hydrographie—Biologisches* Supplement 4:1–125.

Thompson, P. L., L. M. Guzman, L. De Meester, Z. Horváth, R. Ptacnik, B. Vanschoenwinkel, D. S. Viana, and J. M. Chase. 2020. "A Process-Based Metacommunity Framework Linking Local and Regional Scale Community Ecology." *Ecology Letters* 23:1314–29.

Thompson, R. M., E. D. Edwards, A. R. McIntosh, and C. R. Townsend. 2001. "Allocation of Effort in Stream Food-Web Studies: The Best Compromise?" *Marine and Freshwater Research* 52:339–45.

Thorp, J. H., and M. D. Delong. 1994. "The Riverine Productivity Model: An Heuristic View of Carbon Sources and Organic Processing in Large River Ecosystems." *Oikos* 70:305–8.

Thorp, J. H., M. C. Thoms, and M. D. Delong. 2006. "The Riverine Ecosystem Synthesis: Biocomplexity in River Networks across Space and Time." *River Research and Applications* 22:123–47.

Tickner, D., J. J. Opperman, R. Abell, M. Acreman, A. H. Arthington, S. E. Bunn, S. J. Cooke et al. 2020. "Bending the Curve of Global Freshwater Biodiversity Loss: An Emergency Recovery Plan." *BioScience* 70:330–42.

Tiegs, S. D., D. M. Costello, M. W. Isken, G. Woodward, P. B. McIntyre, M. O. Gessner, E. Chauvet et al. 2019. "Global Patterns and Drivers of Ecosystem Functioning in Rivers and Riparian Zones." *Science Advances* 5:eaav0486.

Tierno de Figueroa, J. M., M. J. López-Rodríguez, and M. Villar-Argaiz. 2019. "Spatial and Seasonal Variability in the Trophic Role of Aquatic Insects: An Assessment of Functional Feeding Group Applicability." *Freshwater Biology* 64:954–66.

Tilly, L. J. 1968. "Structure and Dynamics of Cone Spring." *Ecological Monographs* 38:169–97.

Tockner, K., F. Malard, and J. V. Ward. 2000. "An Extension of the Flood Pulse Concept." *Hydrological Processes* 14:2861–83.

Tonina, D., and J. M. Buffington. 2011. "Effects of Stream Discharge, Alluvial Depth and Bar Amplitude on Hyporheic Flow in Pool-Riffle Channels." *Water Resources Research* 47.

Tonkin, J. D., F. Altermatt, D. S. Finn, J. Heino, J. D. Olden, S. U. Pauls, and D. A. Lytle. 2018. "The Role of Dispersal in River Network Metacommunities: Patterns, Processes, and Pathways." *Freshwater Biology* 63:141–63.

Tonkin, J. D., M. T. Bogan, N. Bonada, B. Rios-Touma, and D. A. Lytle. 2017. "Seasonality and Predictability Shape Temporal Species Diversity." *Ecology* 98:1201–16.

Tonkin, J. D., J. D. Olden, D. M. Merritt, L. V. Reynolds, J. S. Rogosch, and D. A. Lytle. 2021. "Designing Flow Regimes to Support Entire River Ecosystems." *Frontiers in Ecology and the Environment* 19:326–33.

Tonkin, J. D., N. L. Poff, N. R. Bond, A. Horne, D. M. Merritt, L. V. Reynolds, J. D. Olden, A. Ruhi, and D. A. Lytle. 2019. "Prepare River Ecosystems for an Uncertain Future." *Nature* 570:301–3.

Torgersen, C. E., C. Le Pichon, A. H. Fullerton, S. J. Dugdale, J. J. Duda, F. Giovannini, É. Tales et al. 2022. "Riverscape Approaches in Practice: Perspectives and Applications." *Biological Reviews* 97:481–504.

Toth, J. 1963. "A Theoretical Analysis of Groundwater Flow in Small Drainage Basins." *Journal of Geophysical Research* 68:4795–812.

Townsend, C. R. 1989. "The Patch Dynamics Concept of Stream Community Ecology." *Journal of the North American Benthological Society* 8:36–50.

Townsend, C. R. 1996. "Concepts in River Ecology: Pattern and Process in the Catchment Hierarchy." *Archiv für Hydrobiologie* Supplement 113:3–21.

Townsend, C. R., and A. G. Hildrew. 1976. "Field Experiments on the Drifting, Colonisation and Continuous Redistribution of Stream Benthos." *Journal of Animal Ecology* 45:759–72.

Townsend, C. R., and A. G. Hildrew. 1994. "Species Traits in Relation to a Habitat Templet for River Systems." *Freshwater Biology* 31:265–75.

Townsend, C. R., M. R. Scarsbrook, and S. Dolédec. 1997. "The Intermediate Disturbance Hypothesis, Refugia, and Biodiversity in Streams." *Limnology and Oceanography* 42:938–49.

Traugott, M., B. Thalinger, C. Wallinger, and D. Sint. 2021. "Fish as Predators and Prey: DNA-Based Assessment of Their Role in Food Webs." *Journal of Fish Biology* 98:367–82.

Triska, F. J., V. C. Kennedy, R. J. Avanzino, G. W. Zellweger, and K. E. Bencala. 1989a. "Retention and Transport of Nutrients in a Third-Order Stream: Channel Processes." *Ecology* 70:1877–92.

Triska, F. J., V. C. Kennedy, R. J. Avanzino, G. W. Zellweger, and K. E. Bencala. 1989b. "Retention and Transport of Nutrients in a Third-Order Stream in Northwestern California: Hyporheic Processes." *Ecology* 70:1893–905.

Triska, F. J., and J. R. Sedell. 1976. "Decomposition of 4 Species of Leaf Litter in Response to Nitrate Manipulation." *Ecology* 57:783–92.

Triska, F. J., J. R. Sedell, K. Cromack, S. V. Gregory, and F. M. McCorison. 1984. "Nitrogen Budget for a Small Coniferous Forest Stream." *Ecological Monographs* 54:119–40.

Trush, W. J. 1991. "The Influence of Channel Morphology on Spawning Steelhead Trout in South Fork Eel River Tributaries." PhD dissertation, University of California, Berkeley.

Twining, C. W., J. T. Brenna, P. Lawrence, D. W. Winkler, A. S. Flecker, and N. G. Hairston Jr. 2019. "Aquatic and Terrestrial Resources Are Not Nutritionally Reciprocal for Consumers." *Functional Ecology* 33:2042–52.

Uehlinger, U. 2000. "Resistance and Resilience of Ecosystem Metabolism in a Flood-Prone River System." *Freshwater Biology* 45:319–32.

Uehlinger, U., and M. W. Naegeli. 1998. "Ecosystem Metabolism, Disturbance, and Stability in a Prealpine Gravel Bed River." *Journal of the North American Benthological Society* 17:165–78.

Uno, H. 2016. "Stream Thermal Heterogeneity Prolongs Aquatic-Terrestrial Subsidy and Enhances Riparian Spider Growth." *Ecology* 97:2547–53.

Uno, H., and M. E. Power. 2015. "Mainstem-Tributary Linkages by Mayfly Migration Help Sustain Salmonids in a Warming River Network." *Ecology Letters* 18:1012–20.

Usseglio-Polatera, P., M. Bournaud, P. Richoux, and H. Tachet. 2000. "Biological and Ecological Traits of Benthic Freshwater Macroinvertebrates: Relationships and Definition of Groups with Similar Traits." *Freshwater Biology* 43:175–205.

Vadeboncoeur, Y., and M. E. Power. 2017. "Attached Algae: The Cryptic Base of Inverted Trophic Pyramids in Freshwaters." *Annual Review of Ecology, Evolution, and Systematics* 48:255–79.

Valett, H. M., C. L. Crenshaw, and P. F. Wagner. 2002. "Stream Nutrient Uptake, Forest Succession and Biogeochemical Theory." *Ecology* 83:2888–901.

Valett, H. M., S. G. Fisher, N. B. Grimm, and P. Camill. 1994. "Vertical Hydrologic Exchange and Ecological Stability of a Desert Stream Ecosystem." *Ecology* 75:548–60.

Valett, H. M., S. G. Fisher, and E. H. Stanley. 1990. "Physical and Chemical Characteristics of the Hyporheic Zone of a Sonoran Desert Stream." *Journal of the North American Benthological Society* 9:201 15.

Vannote, R. L., G. W. Minshall, K. W. Cummins, J. R. Sedell, and C. E. Cushing. 1980. "The River Continuum Concept." *Canadian Journal of Fisheries and Aquatic Sciences* 37:130–37.

Vellend, M. 2010. "Conceptual Synthesis in Community Ecology." *Quarterly Review of Biology* 85:183–206.

Venarsky, M. P., J. P. Benstead, A. D. Huryn, B. M. Huntsman, J. W. Edmonds, R. H. Findlay, and J. B. Wallace. 2018. "Experimental Detritus Manipulations Unite Surface and Cave Stream Ecosystems along a Common Energy Gradient." *Ecosystems* 21:629–42.

Vitousek, P. M., and W. A. Reiners. 1975. "Ecosystem Succession and Nutrient Retention: A Hypothesis." *BioScience* 25:376–81.

Wallace, J. B., S. L. Eggert, J. L. Meyer, and J. R. Webster. 1997. "Multiple Trophic Levels of a Forest Stream Linked to Terrestrial Litter Inputs." *Science* 277:102–4.

Wallace, J. B., S. L. Eggert, J. L. Meyer, and J. R. Webster. 1999. "Effects of Resource Limitation on a Detrital Based Ecosystem." *Ecological Monographs* 69:409–42.

Wallace, J. B., S. L. Eggert, J. L. Meyer, and J. R. Webster. 2015. "Stream Invertebrate Productivity Linked to Forest Subsidies: 37 Stream-Years of Reference and Experimental Data." *Ecology* 96:1213–28.

Wallace, J. B., and J. R. Webster. 1996. "The Role of Macroinvertebrates in Stream Ecosystem Function." *Annual Review of Entomology* 41:115–39.

Wallace, J. B., J. R. Webster, and T. F. Cuffney. 1982. "Stream Detritus Dynamics: Regulation by Invertebrate Consumers." *Oecologia* 53:197–200.

Wallace, J. B., J. R. Webster, and W. R. Woodall. 1977. "The Role of Filter Feeders in Flowing Waters." *Archiv für Hydrobiologie* 79:506–32.

Ward, J. V. 1989. "The Four-Dimensional Nature of Lotic Ecosystems." *Journal of the North American Benthological Society* 8:2–8.

Ward, J. V. 1998. "Riverine Landscapes: Biodiversity Patterns, Disturbance Regimes, and Aquatic Conservation." *Biological Conservation* 83:269–78.

Ward, J. V., and J. A. Stanford. 1983. "The Serial Discontinuity Concept of Lotic Ecosystems." In *Dynamics of Lotic Ecosystems*, edited by T. D. Fontaine and S. M. Bartell, 29–42. Ann Arbor, MI: Ann Arbor Science.

Ward, J. V., and J. A. Stanford. 1995a. "The Serial Discontinuity Concept: Extending the Model to Floodplain Rivers." *Regulated Rivers: Research & Management* 10:159–68.

Ward, J. V., and J. A. Stanford. 1995b. "Ecological Connectivity in Alluvial River Ecosystems and Its Disruption by Flow Regulation." *Regulated Rivers: Research & Management* 11:105–19.

Warren, C. E., J. H. Wales, G. E. Davis, and P. Doudoroff. 1964. "Trout Production in an Experimental Stream Enriched with Sucrose." *Journal of Wildlife Management* 28:617–60.

Waters, T. F. 1961. "Standing Crop and Drift of Stream Bottom Organisms." *Ecology* 42:532–37.

Waters, T. F. 1962. "Diurnal Periodicity in the Drift of Stream Invertebrates." *Ecology* 43:316–20.

Waters, T. F. 1964. "Recolonization of Denuded Stream Bottom Areas by Drift." *Transactions of the American Fisheries Society* 93:311–15.

Waters, T. F. 1965. "Interpretation of Invertebrate Drift in Streams." *Ecology* 46:327–34.

Waters, T. F. 1966. "Production Rate, Population Density, and Drift of a Stream Invertebrate." *Ecology* 47:595–604.

Waters, T. F. 1988. "Fish Production–Benthos Production Relationships in Trout Streams." *Polskie Archiwum Hydrobiologii* 35:545–61.

Webster, J. R., and E. F. Benfield. 1986. "Vascular Plant Breakdown in Freshwater Ecosystems." *Annual Review of Ecology and Systematics* 17:567–94.

Webster, J. R., and J. L. Meyer. 1997. "Organic Matter Budgets for Streams: A Synthesis." *Journal of the North American Benthological Society* 16:141–61.

Webster, J. R., and B. C. Patten. 1979. "Effects of Watershed Perturbation on Stream Potassium and Calcium Dynamics." *Ecological Monographs* 49:51–72.

Welcomme, R. L. 1985. "River Fisheries." FAO Technical Paper 262. Rome: Food and Agricultural Organization of the United Nations.

Wetzel, R. G. 1995. "Death, Detritus, and Energy Flow in Aquatic Ecosystems." *Freshwater Biology* 33:83–89.

Whitehead, H. 1935. "An Ecological Study of the Invertebrate Fauna of a Chalk Stream near Great Driffield, Yorkshire." *Journal of Animal Ecology* 4:58–78.

Wiens, J. A. 2002. "Riverine Landscapes: Taking Landscape Ecology into the Water." *Freshwater Biology* 47:501–15.

Wiggins, G. B., and R. J. Mackay. 1978. "Some Relationships between Systematics and Trophic Ecology in Nearctic Aquatic Insects, with Special Reference to Trichoptera." *Ecology* 59:1211–20.

Wilcox, A., B. L. Peckarsky, B. W. Taylor, and A. C. Encalada. 2008. "Hydraulic and Geomorphic Effects on Mayfly Drift in High-Gradient Streams at Moderate Discharges." *Ecohydrology* 1:176–86.

Williams, D. D., and H. B. N. Hynes. 1974. "The Occurrence of Benthos Deep in the Substratum of a Stream." *Freshwater Biology* 4:233–56.

Williams, D. D., and H. B. N. Hynes. 1976. "The Recolonization Mechanisms of Stream Benthos." *Oikos* 27:265–72.

Williamson, T. J., W. F. Cross, J. P. Benstead, G. M. Gíslason, J. M. Hood, A. D. Huryn, P. W. Johnson, and J. R. Welter. 2016. "Warming Alters Coupled Carbon and Nutrient Cycles in Experimental Streams." *Global Change Biology* 22:2152–64.

Willson, M. F., S. M. Gende, and B. H. Marston. 1998. "Fishes and the Forest." *BioScience* 48:455–62.

Willson, M. F., and K. C. Halupka. 1995. "Anadromous Fish as Keystone Species in Vertebrate Communities." *Conservation Biology* 9:489–97.

Winberg, G. 1937. "Beobachtungen uber die Intensitat der Atmung und Photosynthese des Planktons der Fischzuchtteiche." *Arbeiten der Limnologischen Station zu Kossino* 21:61–74.

Winemiller, K. O. 1990. "Spatial and Temporal Variation in Tropical Fish Trophic Networks." *Ecological Monographs* 60:331–36.

Winterbourn, M. J., J. S. Rounick, and B. Cowie. 1981. "Are New Zealand Stream Ecosystems Really Different?" *New Zealand Journal of Marine and Freshwater Research* 15:321–28.

Wipfli, M. S., J. Hudson, and J. Caouette. 1998. "Influence of Salmon Carcasses on Stream Productivity: Response of Biofilm and Benthic Macroinvertebrates in Southeastern Alaska, USA." *Canadian Journal of Fisheries and Aquatic Sciences* 55:1503–11.

Wohl, E. 2010. "A Brief Review of the Process Domain Concept and Its Application to Quantifying Sediment Dynamics in Bedrock Canyons." *Terra Nova* 22:411–16.

Wohl, E. 2014. "A Legacy of Absence: Wood Removal in US Rivers." *Progress in Physical Geography* 38:637–63.

Wohl, E. 2017. "Bridging the Gaps: An Overview of Wood across Time and Space in Diverse Rivers." *Geomorphology* 279:3–26.

Wohl, E. 2020. "Wood Process Domains and Wood Loads on Floodplains." *Earth Surface Processes and Landforms* 45:144–56.

Wohl, E. 2021. "Legacy Effects of Loss of Beavers in the Continental United States." *Environmental Research Letters* 16:025010.

Wohl, E., P. L. Angermeier, B. Bledsoe, G. M. Kondolf, L. MacDonnell, D. M. Merritt, M. A. Palmer, N. L. Poff, and D. Tarboton. 2005. "River Restoration." *Water Resources Research* 41:W10301.

Wollheim, W. M., T. K. Harms, A. L. Robison, L. E. Koenig, A. M. Helton, C. Song, W. B. Bowden, and J. C. Finlay. 2022. "Superlinear Scaling of Riverine Biogeochemical Function with Watershed Size." *Nature Communications* 13:1230.

Wolman, M. G., and J. P. Miller. 1960. "Magnitude and Frequency of Forces in Geomorphic Processes." *Journal of Geology* 68:54–74.

Woodward, G., M. O. Gessner, P. S. Giller, V. Gulis, S. Hladyz, A. Lecerf, B. Malmqvist et al. 2012. "Continental-Scale Effects of Nutrient Pollution on Stream Ecosystem Functioning." *Science* 336:1438–40.

Woodward, G., and A. G. Hildrew. 2002. "Food Web Structure in Riverine Landscapes." *Freshwater Biology* 47:777–98.

Woodward, G., D. C. Speirs, and A. G. Hildrew. 2005. "Quantification and Resolution of a Complex, Size-Structured Food Web." *Advances in Ecological Research* 36:85–135.

Woodward, G., R. M. Thompson, C. R. Townsend, and A. G. Hildrew. 2005. "Pattern and Process in Food Webs: Evidence from Running Waters." In *Aquatic Food Webs: An Ecosystem Approach*, edited by A. Belgrano, U. M. Scharler, J. Dunne, and R. E. Ulanowicz, 51–66. Oxford, UK: Oxford University Press.

Woodward, G., and P. H. Warren. 2007. "Body Size and Predatory Interactions in Freshwaters: Scaling from Individuals to Communities." In *Body Size: The Structure and Function of Aquatic Ecosystems*, edited by A. G. Hildrew, D. Raffaelli, and V. Edmonds-Brown, 98–117. Cambridge, UK: Cambridge University Press.

Woodwell, G. M., and R. H. Whittaker. 1968. "Primary Production in Terrestrial Ecosystems." *American Zoologist* 8:19–30.

Wootton, J. T., M. S. Parker, and M. E. Power. 1996. "The Effect of Disturbance on River Food Webs." *Science* 273:1558–60.

Wotton, R. S. 1994. *The Biology of Particles in Aquatic Systems*. Boca Raton, FL: CRC Press.

Wright, J. C. and J. K. Mills. 1967. "Productivity Studies on the Madison River, Yellowstone National Park." *Limnology and Oceanography* 12:568–77.

Wright, J. P., C. G. Jones, and A. S. Flecker. 2002. "An Ecosystem Engineer, the Beaver, Increases Species Richness at the Landscape Scale." *Oecologia* 132:96–101.

Wu, X., X. Zheng, L. Yu, R. Lu, Q. Zhang, X. J. Luo, and B. X. Mai. 2022. "Biomagnification of Persistent Organic Pollutants from Terrestrial and Aquatic

Invertebrates to Songbirds: Associations with Physiochemical and Ecological Indicators." *Environmental Science and Technology* 56:12200–12209.

Young, R. G., and A. D. Huryn. 1998. "Comment: Improvements to the Diurnal Upstream-Downstream Dissolved Oxygen Change Technique for Determining Whole-Stream Metabolism in Small Streams." *Canadian Journal of Fisheries and Aquatic Sciences* 55:1784–85.

Zarfl, C., A. E. Lumsdon, J. Berlekamp, L. Tydecks, and K. Tockner. 2015. "A Global Boom in Hydropower Dam Construction." *Aquatic Sciences* 77:161–70.

Zipper, S. C., J. C. Hammond, M. Shanafield, M. Zimmer, T. Datry, C. N. Jones, K. E. Kaiser, et al. 2021. "Pervasive Changes in Stream Intermittency across the United States." *Environmental Research Letters* 16:p.084033.

Zou, K., E. Thébault, G. Lacroix, and S. Barot. 2016. "Interactions between the Green and Brown Food Web Determine Ecosystem Functioning." *Functional Ecology* 30:1454–65.

# CONTRIBUTORS

*Colden V. Baxter*: Stream Ecology Center, Department of Biological Sciences, Idaho State University, Idaho, USA

*Jonathan P. Benstead*: Department of Biological Sciences, University of Alabama, Alabama, USA

*Emily S. Bernhardt*: Department of Biology, Duke University, North Carolina, USA

*Luz Boyero*: Department of Plant Biology and Ecology, University of the Basque Country (UPV/EHU), Bilbao, Spain

*Wyatt F. Cross*: Department of Ecology, Montana State University, Montana, USA

*Debra S. Finn*: Department of Biology, Missouri State University, Missouri, USA

*Mark O. Gessner*: Leibniz Institute of Freshwater Ecology and Inland Fisheries (IGB), Stechlin, Germany, and Berlin Institute of Technology (TU Berlin), Berlin, Germany

*Robert O. Hall Jr.*: Flathead Lake Biological Station, University of Montana, Montana, USA

*Alan G. Hildrew*: Queen Mary, University of London, UK, and Freshwater Biological Association, Windermere, UK

*Alexander D. Huryn*: Department of Biological Sciences, University of Alabama, Alabama, USA

*Amy M. Marcarelli*: Biological Sciences, Michigan Technological University, Michigan, USA

*Eugènia Martí*: Center for Advanced Studies of Blanes (CEAB-CSIC), Spain

*Angus R. McIntosh*: Te Kura Pūtaiao Koiora, Te Whare Wānanga o Waitaha, Aotearoa | School of Biological Sciences, University of Canterbury, New Zealand

*David M. Merritt*: US Forest Service Research and Development, Colorado, USA

*Barbara L. Peckarsky*: Integrative Biology, University of Wisconsin–Madison, USA, and the Rocky Mountain Biological Laboratory, Colorado, USA

*Marc Peipoch*: Stroud Water Research Center, Pennsylvania, USA

*Geoffrey C. Poole*: Department of Land Resources and Environmental Sciences, Montana State University, Montana, USA

*Mary E. Power*: Integrative Biology, University of California, Berkeley, USA, and the Angelo Coast Range Reserve, California, USA

*Stephen Rice*: Geography and Environment, Loughborough University, Loughborough, UK

*Ryan A. Sponseller*: Department of Ecology and Environmental Science, Umeå University, Umeå, Sweden

*Amanda Subalusky*: Department of Biology, University of Florida, Florida, USA

*Jennifer L. Tank*: Department of Biological Sciences, University of Notre Dame, Indiana, USA

*Jonathan D. Tonkin*: Te Kura Pūtaiao Koiora, Te Whare Wānanga o Waitaha, Aotearoa | School of Biological Sciences, University of Canterbury, New Zealand

*Hiromi Uno*: Faculty of Environmental Earth Science, Hokkaido University, Hokkaido, Japan

*H. Maurice Valett*: Division of Biological Sciences, University of Montana, Montana, USA

*Daniel von Schiller*: Department of Evolutionary Biology, Ecology and Environmental Sciences, University of Barcelona, Spain

*Ellen Wohl*: Department of Geosciences, Colorado State University, Colorado, USA

# INDEX

The letter *f* following a page number denotes a figure.